I shall fear not, for it is a good day to die. That's what I said to myself as I withstood all of the psychological and emotional might Exxonmobil's demons and witches attempted to provide.

It got to the point where I offered to meet up for a gun fight with more the ten cowardice Exxonmobil employees. I would call in a complaint of hazardous air inside of Exxonmobil to Texas Commission on Environmental Quality, and when TCEQ got there I would say, "TCEQ is at the gate—I'm the motherfucker that called them—if you don't like that, go get your pistol, meet me down the road, and we will settle it."

Their strength came from Satan, which isn't strength at all. They have not any rights in heaven, nor on earth. We have authority over Satan, demons, and the Sabbath.

I literally scared the hell out of them, and continued as I chose. I winced at their ignoble cowardliness, and told everybody—like a crazy person—what I was experiencing, and what was happening behind the fence. A friend suggested I write a pamphlet and give it out like Thomas Pain. I wrote the exposé "Behind the Fence" because I couldn't tell the story anymore, plus everyone was exhausted by me and my intensity.

BEHIND *the* FENCE

Exposé on Refinery Practices

Behind the Fence

Part 1: The Root Cause

I would go to unit CHD-1 and an operator – who was only a few years from retirement – would hassle me about a sump that was emitting dangerous fumes. This 36" manway opening had been exposed to the atmosphere since hurricane Ike. Management would ignore all complaints from me and the operator. Management would have plastic sheeting covering this opening and weigh it down with sand bags, which is not an effective way to capture fumes. I would constantly remind and work with them to provide solutions – even if it was not accepted. Optimistically and naively it went on too long. After dealing with these people long enough, you know what they are going to do before they know what they are going to do.

My conscience, my body, and my soul could take no more. So I contacted Texas Commission on Environmental Quality.

The next few months were extremely painful. Verbal abuse and attempts to establish dominance were relayed to me – all unsuccessful.

Soon, the State of Texas began their investigations. A few months later, management was eager to learn of any concerns I had.

Here are the root causes of my concerns:

A

We have an epidemic of cognitive dissonance. Mental stress and discomfort is experienced by the people who work in these facilities. Thes people hold two or more contradictory beliefs and values at the same time - and are confronted by new information that conflicts with existing beliefs, ideas, or values. These people will actively avoid situations and new information that will likely increase their discomfort.

B

Refining oil is no more profitable than the Sabine River Authority pumping water as a service. It's a break-even after you give your percentage for the privilege of using whatever corporate name you subscribe to. Most money is syphoned off at the corporate level – little is put back into maintenance.

Since the United States and Iran have newly entered the oil exporting game, the price of oil has plummeted. Saudi Arabia is selling a 1/3 of its holdings and production cheap. OPEC and Russia have the ability to drip oil from here to eternity, causing the supply to be greater than the

mand- hence the decline of fracking and exploration. There is no need
r exploration because there is already a surplus in the "free market"- if
ere ever was a free market. And hopefully more supply will come.

The truth is, all refineries would run at 100% even if oil was $2.00 a
rrel – because there is a need. But what is the true cost?

T

ExxonMobil Refinery Beaumont is considered a heritage site (from
e Mobil days), the Exxon portion of that name treats this plant like an
wanted step child – giving a very small budget to maintain this facility.
e expectation is to continue to be profitable, but what they are really
ying for is a ticking time bomb.

A

The refinery and chemical plant in Beaumont are being run into
e ground by the business teams. The only subject that matters to them
"the bottom line" - and the very infrastructures of these facilities are a
nger to our community and the lives within and around them.

Instead of maintaining equipment properly, the business teams
der Band-Aid upon Band-Aid in order to cut costs and prove to the
areholders that it is still profitable. Exxon's policy is "run to failure" -
at means, something would have to catch fire or explode before they
ould fix their equipment. However, they spent over a million dollars to
ess-up the parade procession for the shareholders to view the century-
d decaying carcass. These practices put plant workers lives and our
mmunity in danger - not only physically but also economically.

On February 18, 2015, at an ExxonMobil refinery in Torrance,
alifornia an explosion occurred due to "run to failure" policy. The FCC
luid Catalytic Cracking) unit shut down and steam was forced into
e reactor to prevent hydrocarbons from flowing back from the main
stillation column. Instead steam escaped through an open flange on the
pander – preventing operations from continuing repairs. It had traveled
rough a leaking slide valve connected to the reactor. Supervisors
duced the amount of steam so work could continue. The pressure of
e steam reduced and hydrocarbons flowed into the column – finding the
nition source and exploding.

The U.S. Chemical Safety Board uncovered multiple process safety
anagement deficiencies that led to the explosion. Federal and State
egulators have said ExxonMobil deliberately failed to fix equipment it
new could lead to a life-threatening explosion.

Soon after, gas prices soared to $5 a gallon in Los Angeles.

PBF, a petroleum refiner and supplier of unbranded transportation fuel – was in the process of buying this facility when the explosion occurred. PBF was also contracted to purchase the Chalmette refinery in Louisiana, through a separate and independent bidding process, in which ExxonMobil holds 50% interest.

E

Safety is an illusion and its only purpose is to minimize provable personal injury lawsuits. It doesn't matter how many Loss Prevention Self Assessments (LPSA) occur when complaints of being poisoned by benzene and hydrogen sulfide are pushed to the side. You can be thrown out of the plant for not having your gloves on your person – but you can breathe all the benzene you want on the house. If you get injured – Exxon philosophy is, "you must have not done your LPSA." Upper management made people break procedure in the past and the outcome was death – all to save a dime. The contractors could have stopped the job – but the fear not having a pay check is greater than the fear of death.

C

Some refineries are union based and its purpose is to create a barrier against corporate tyranny. And where is the representation when safety issues occur? If the union can't help its members with these issues then they are nothing more than a collection agency – absentee land lord. It appears that they can only keep their members in a job and that's about it. And some of them don't deserve it. This must be why less than 7% of private sector workers have unions and inequality of wages is on the rise. is also said that the union is only as good as its members - that resonates as well.

Z

Make no mistake - the operators have an ownership of this debacle - sabotaging the very lifesaving controls that are put in place by law for their protection. The controls I speak of are carbon filters and flame arresters. Carbon filters are designed to capture emissions and decrease exposure to toxic fumes, and are bypassed when possible because of cost. Flame arresters are designed to extinguish burning hoses not to be a good access point to clog inlet hoses to carbon filters. Is it that they are burdened by bureaucracy or lazy and ignorant?

The environmental controls are viewed as a "cover your ass" situation – as opposed to the actual safeguards they are designed to be, n more valuable to them than a box checked for government records. Trying to educate them in the matter is futile – because management has already trained the operators to the only reality they have time to accept. "The definition of insanity is doing the same thing over and over and expecting fferent results." Albert Einstein. It's acceptances of insanity that breeds complacency.

What's the advantage?

𝓗

Management cannot move up if they spend money in their unit. The thought process is, "Convince your subordinate it is detrimental to their livelihood to spend money or you will have trouble receiving your bonus." The operator's explanation is usually, "But I'm in here" – meaning they have been hiding from the smell inside and would mitigate their exposure as much as allowed. Most times management stays only two years in a job. By the time their screw-up hits the fan, they are already in another job in another state. It's called migratory management; someone else will reap what the last employee sows. That's the corporate way.

Θ

You can no longer make a fortune in shift work. That time died in the 90's – when they forced anyone who made a fortune to retire. The new generation received much less and the precedent was set. These plans take time to develop and execute. And we are looking at the outcome now. A 2% raise doesn't stand equal with the cost of living.

I

Operators are put in a position of cannibalization - corporate psychopaths manipulating their employees to back-bite, or to brown-nose. Its only purpose is to keep their minds occupied from looking at or having time to deal with real issues. Keeping them divided is the best way to conquer. The only way to move up in the company is to suck up or screw up - to be so incompetent that management takes you away from machinery and puts you in charge of working men. Because it is unsafe to have a screw up on the unit. It reminds me of the government – where the right hand doesn't know what the left hand is doing. Zero cooperation within these establishments.

Let's take a minute to understand what "Corporate Psychopath" means. The psychopath is one of the most studied personality disorders. There are 20 well documented characteristics that form this unique human personality syndrome. At first glance, they are impressive, charming, and entertaining - eager to impress with accomplishments which can result in instant wide rapport. Under the mask of success is a liar who cons, manipulates, and deceives for self-interest. They become bored easily and seek stimulation from mind games with strong desire to win. The corporate psychopath is driven by their victims' vulnerability - a perverted pleasure from hunting and abusing victims. Corporate Psychopaths lack basic human emotions and feelings of guilt, remorse, or empathy. They mimic others emotional response and it is visible in their shallow sentiment. The irresponsible and impulsive psychopath lives a parasitic and predatory lifestyle - using people for money, food, shelter, sex, power, and influence. This can all be found on the Federal Bureau of Investigation (FBI Law Enforcement Bulletin) – called The Corporate Psychopath – by Paul Babiak, Ph.D., and Mary Ellen O'Toole, Ph.D. It is important to identify and acknowledge a psychopath so that their hunting grounds are tarnished.

$$I\alpha$$

It should not be acceptable to openly brag about pushing the environmental team aside or shutting down their internal investigation because of a call that was made to their supervisor. People with shrunken pineal glands due to over exposure view environmentalism as tree hugging – when its intention is to protect all life in and around these facilities.

$$I\beta$$

Engineers pay a high price and debt interest owed on the education they have received. The truth is, they have to forget almost everything they learned in school and are retrained by someone who has 30 years of experience without an education and gets paid less. Experience trumps the equation. These kids paid a university so they can skip manual labor. They pay these entities to agree and vouch for them by trade of debt, and now they live only on paper or computer screen. They have no idea what it takes to execute task or the true outcome and repercussions of their actions. By the time they learn their job the two year migration starts, and they are off to be educated by some other old plant rat. It takes time to gain wisdom. And without experience and wisdom you are not fit to be in charge of working men. Resentment has grown for the same reason as before – divide.

$$I\gamma$$

One of the things keeping the plants alive are contractors, and they are viewed as a disease – even though most on staff employees were formerly contractors. Contractors have the only true craft left in this market and do most of the physical work. Operators' exposure to toxic chemicals release high amounts of cortisol and cytokines that rupture the blood brain barrier which also accounts for their short term memory. They forgot how it feels to work hard. ExxonMobil Refinery Beaumont usually kills one contractor per shut down. That's at least one contractor every two years, and they continue to meet their quota.

Most of these companies are understaffed and use recommendation 755: Fatigue Risk Management Systems for Personnel the Refining and Petrochemical Industries - brought to you by the Petroleum Institute.

Recommendation 755 is a fatigue policy used at the whim of management. When the Steel Workers Union , who at one time partnered with the Petroleum Institute in this endeavor, saw that the Petroleum Institute was headed down a one way street – the union backed out of the deal. And the Petroleum Institute went full steam ahead.

The U.S. Chemical Safety Board criticized this policy because it was not a result of an effective consensus process, it didn't provide firm requirements for employers in essential elements, but rather suggested options to consider. It unreasonably emphasizes personal components of fatigue prevention – such as allowing self-evaluation, evaluations by supervisors, and educating employees about fatigue without objective evidence to support these components. And its limits on hours and days at work are more permissive and less protective than suggested by the U.S. Chemical Safety Board. In layman's terms - try to get two from one - greed the driver.

Why must it be so difficult?

Why must this corporate state be a constant source of pain and anxiety?

It's all the same oil being refined by the same companies – going down the same Colonial Pipeline. There is no difference between Shell gas, Texaco gas, or ExxonMobil gas – it all goes to the same holding facilities. There is no difference between Pennzoil, Quaker State, or any other motor oil. It's all bottled at the same blending and packaging facility.

Refining oil is a moot point. Even ExxonMobil doesn't want to do it, or they wouldn't be selling their plants. And we will continue to use these fuels as long as there is no alternative.

Why should we continue to put our lives and the livelihood of our community in danger for the bottom line profit when there is not that much profit to begin with?

I do not want for Southeast Texas and Southwest Louisiana to be the toxic sacrificial lamb for the rest of the United States. I would like BF or someone equivalent to purchase ExxonMobil Refinery before ExxonMobil lays waste to its equipment with their locust ways, devouring everything in sight.

These corporate psychopaths try to convince you that they can just keep their money in the banks and have better dividends as opposed to maintaining their facilities correctly. Everyone knows that a tangible asset is more valuable than a fiat number on a spread sheet readily available to be taken away from them at the next and near manufactured "economical down turn." And if they don't know that – then they have more money than they have sense.

Let's review Leverage, verb – to use something to maximum advantage. When you think of union contract, the next word in your mind will be leverage. You have now learned that the corporate psychopath does not care about your health or well-being. So what is there to do to gain leverage you might ask? All plants and refineries are hiding environmental and safety violations because of cost. It is time to start gathering and journaling these violations. Texas Commission on Environmental Quality (TCEQ) and Occupation Safety and Health Administration (OSHA) under the Department of Labor have the authority to protect you, but do not have any power behind them. Their teeth have been filed down because you the reader is unaware of the inner workings and do not have a working relationship with them. Their power comes from you, and public opinion. You are unaware of the lax federal regulations because you have not taken the time to investigate. You have not taken the time to contact your elected representatives to talk with them about your concerns. Mostly because you fear and distrust your representatives, you fear losing your job, you know that world corruption is even in our little town. Only when you can't take it anymore will you make a change. And I believe we have hit the high water mark. Challenge the beast if you dare, if you have any honor or courage. Call and force these government entities to look at these violations. By the time your contract comes around, they will be eager to sign. If they have any arrogance left in them, you didn't cause them enough pain and anxiety. If these government entities do not satisfy our needs, then we will place more focus on them.

The United States is working on a partnership with eleven other countries. The Transatlantic Trade and Investment Partnership (TTIP) would allow corporations to sue governments for impeding on their profits. Meaning Phillip Morris would be able to sue the United States for blocking cigarette ads that target young people and children, and they are doing this in Uruguay, Norway, and Australia. ExxonMobil would be able to release emissions legally if they could prove they were losing profit. A free society means no one aggresses upon anyone's person or property. Every person has the absolute right of property in their own self – that is bestowed to us by God. And if these companies aggress upon our person or property then we should be able to sue them for hindering our God given right to be free.

We can no longer operate in this oligarchic corporate state. Trying to reconfigure the global economy in the form of neo feudalism will back fire. We are not serfs – we are humans. The only difference between wage

bor and slavery is that it is temporary. We cannot sustain ourselves
competing with prison labor in China. It is all unnecessary in the
velopment of the human race. Corporate nationalism will hollow us from
e inside out – one and all. If corporate conglomerates cannot justify
emselves then they are null and void.

Cancer rates in Southeast Texas are high to say the least. In 2015,
,520 people died from cancer in Texas. More than 9,500 were lung and
onchus related - and more than 113,000 new cases of cancer appeared
2015 alone. This information can be found from the American Cancer
ociety. We are well above the national average. This is the last thing I want
y community to be known for.

Margaret Fuller wrote in the New York Daily Tribune a story
led "Fourth of July" in 1845. She berated America for its tolerance of
avery in their mean pursuit of wealth. Fuller noted that private lives,
ore than public measure, are the salvation of our country. She called
r "individuals" who could be shining examples and whose deep rooted
aracters could not be moved by flattery, by fear, even by hope, for they
ly work in faith - to awaken their neighbors from their torpid lives of
pediency to lives of principle. We are in a time of need for people with
ep roots who cannot be moved anything but faith, people who will shine
eir light into darkness –dissipating it quickly.

ExxonMobil has requested that I be removed from the Coker site.
ey say it's a conflict of interest for me to speak to the operators at said
it - mostly because I give them valuable information that will protect their
es. The true conflict is the stifling of this information and ExxonMobil's
illingness to slowly and unnecessary poisoning of their employees. It is
t time to look the other way. It's time to come clean. Release that weight
f of your soul.

I write this exposé with love for my friends and family that work at
fineries and chemical plants and are subjected to unnecessary practices
d to the families who's loved ones were killed while working in these
cilities. And I write with love for my friends and family who are professors
d students at Lamar University – who occasionally have to shelter in
ace because of its close proximity. I call out to you in the industry, and
ea to everyone who is germane to the subject matter. Reconsider your
actices before a catastrophe occurs, and our community is crippled
hysically, emotionally, and economically.

Nathaniel Welch

May 5, 2016

I gave the exposé to everyone including people I liked and didn't like in the refinery. One man thought it was an ass kissing piece of literature and was stunned silent when he realized it was exposure. I told him I couldn't even take a picture of this priceless moment. Then I pointed to my heart and told him I was saving that look on his face right here.

After the exposé, Exxonmobil deactivated my entry access—they wouldn't dare confront me face to face. My company offered me a job in Houston with same pay. I wasn't giving their corporate psychopath the opportunity to find a reason to cause me trouble. By then, I had a gut full of corporate life and laid in the bed for a long while.

I gave the exposé to many major news outlets—they took what I said and used it as their own. The message was out, so who fucked who?

I still had to pay bills, so I reverted back to what my family taught me—carpentry. I did a lot of handyman work like build sheds, paint houses, clean toilets—whatever I had to do to survive.

Southeast Texas had a major flood—some kinfolk had to rescue my mother from about six feet of bayou water. I did a lot of free work for people who had nothing and lost the rest.

While neck deep in the destruction, I noticed more than fifty people died unexpectedly—some of which were healthy athletic people. It was from pollution—from what I thought to be from the refineries and chemical plants in the area. Don't get me wrong

—refineries and chemical plants in Southeast Texas pollute. I got so sick that I expected to die—obviously I survived.

A friend—very forcefully—made me get out of Southeast Texas before I did die from pollution, so I moved to Austin. It was there that I realized Austin was over-populated and that the whole world was polluted to the point where people were going to die in mass quantities soon. As calmly as I could—which wasn't very calm at all—I explained to everyone I encountered what was to come. There was nowhere to run anymore.

People began to understand they couldn't walk across a parking lot because the pollution replaced their breathing air while previously thinking it was the heat causing their exhaustion—powers and principalities couldn't have people understanding their demise.

I knew in the spring of 2019 that people were going to die from pollution from Africa and Mexico in the summer. They had to have a cover story explaining why people were dying. I told everyone who would hear that 2020 will be the beginning of clear vision—just as 20/20 indicates clear vision.

Hong Kong fought and ran the Red Army back to Beijing, which was the fuse that lit the powder keg on the all-in attack from the communists. I knew what I was looking at—the death of communism.

Communists China barricaded their citizen in their homes and there were lots of reports that peoples bodily motor function shut down and the communist police carried them away like sacks of trash. In Austin, peoples bodies started seizing up and became inflamed in large numbers as the 5G took full effect.

I also knew they were going to dump the stock market and I told people to bet the short two week before the Covid lockdown. I did not bet the short—I couldn't bet against humanity. I had a cash account ready for the drop—the criminals that control the market were only two years late, but now they had a reason.

As seen on TV, Johnson and Johnson was sued and paid billions with only the first five lawsuits—which told me they needed a bailout. I was ready for the fuckery.

As a carpenter, I couldn't go one week without work, nor could I stand the communism in Austin—so I brought my ass back to the bayou and wrote this book.

"The cause of America is in great measure the cause of all mankind". - Thomas Paine

The Kingdom of Saudi

XIV

The Kingdom of Saudi is forsaken and desolate, bereft of the presence of Allah (God-The Source), much like Central and South America. They are the sons of Ishmael (the father of the Arab nation)— nephews to Issac.

It exceeds all other nations in cruelty, and the dead far out number the living. Humans are weary of life and have ceased seeing the universe as worthy of reverent wonder.

Darkness is preferred to light, and not many eyes rise to heaven. The pure are thought insane and the impure honored as wise. The madman, brave; and the wicked, good.

Sixty nine miles south of Ur was a marshland drained by oil exploitation. The Ma'Dans were driven from their homeland, and the largest wetland eco-system in the Middle East was annihilated.

I'll bet you didn't know there was a wetland in such a place. It was only 17,000 sq kilometers (6,500 square miles), about the size of Wales. Now only 10 percent survives. It nourished the cradle of human life, and now the criminally insane bites the hand that feeds them.

Saudi Aramco is the Kingdom of Saudi. It is also the largest company in the world, revenuing over

182 billion (that we know of); not counting human trafficking and other sources of income which are confidential.

They own Standard Oil, Texaco, Exxonmobil, Shell, Total Refining, Motivia, King Salaman Global Maritime Industries Complex, and a great number of other chemical and refining companies on the planet.

They produce 259.9 billion barrels a day and also produce the highest levels of CO2 emissions, replacing the oxygen and nitrogen levels required for human and animal existence, with a towering 1,707 million tons in 2013 alone.

In Part 1 section 14 of "Behind the Fence" I stated that Exxonmobil doesn't want to refine oil based on the selling of their refineries. I didn't realize the Kingdom of Saudi owns controlling shares of most all refineries and chemical plants globally.

The moot point is that there is an oil market at all. The Kingdom of Saudi has a monopoly over the entire market. Trading in this market is fractural fraud in every way. Anti - Trust laws broke up the oil company monopolies in America but couldn't or wouldn't do anything about the Kingdom of Saudi.

They would also make you believe you are in competition with each other for wages. Nothing could be further from the truth. A monopoly means everything involved is monopolized, including human commodities.

Don't kid your self. The cruelty of Capitalism says that because of supply and demand that any product - including human labor—is more valuable if there is a perceived finite supply or less valuable if there is an influx of human labor. Most humans—from birth—have been conditioned to value not themselves.

The people of Orange County, Texas would drink gasoline for $35 an hour, not fully understanding the obscene amount of money the companies that poison them possess. For if they did—they would drag the profiteers and money changers from their holes and burn them alive.

These people are secretly programmed under the disguise of hearing exams and children's news channels in elementary and junior high school—hypnotized to be enslaved physically, sexually, mentally, and emotionally.

Being poor isn't a virtue. You must love yourself before you can love your neighbor.

Silicon Valley is owned by the Kingdom of Saudi and the Communists Party—who all become rich through state sponsored murder, and slavery. Everyone who complies with the Kingdom of Saudi and the Communists Party becomes very rich. You can't do business without their monetary presents. Saudi and communists money is behind all the big startup tech companies, which include Lyft, Uber, Magic Leap and many more name brands.

SoftBank's Vision Fund received $45 billion in Saudi money. This was SoftBank's Vision Fund second payment from the Kingdom of Saudi and one of the largest venture funds to date. The primary Saudi investors are Saudi Aramco Energy Ventures, Saudi Public Investment Fund, Riyadh Valley Company, The Kingdom Holding Company, Vision Venture Capital, and Saad AlSogair.

Most venture capitalists have a vice clause. No investments in drugs, weapons, alcohol, pornography, and, coming soon, dictatorships. The United States Government should follow suit.

The Kingdom of Saudi is looking to launder their blood money in technology and uses technology to further imprison humans—money they got through state sponsored fraud, the degradation of its people, and murder.

In Yemen, the Saudi-led coalition war planes that killed 40 boys (ages 6 to 11) and 11 adults on a school trip were sold by Lockheed Martin to Riyadh Valley Company. The school bus bombing led to 79 people wounded, 56 of them were children—one of the many defilements committed by these warlords.

Intelligence, refueling, and weapons come from the United States and the United Kingdom. In 2015, the state department approved of such measures and the commander-in-chief bragged about how good they were getting at drone strikes. The United States Department of Defense killed thousands to eliminate a few. Drone strikes are cowardice—if you have not the guts to look them in the face, then you shouldn't do it. A video game killing lacks honor in every way.

The Department of Defense says they don't make tactical decisions for the Kingdom of Saudi. But they damn sure supply necessary tools for carnage against civilians. More than $115 billion in weapons were sold to them over this 4 year administration term. They are the United States' largest arms customers and its economy relies on their state sponsored violence for sustenance.

Even Pat Roberts, a television evangelist who I watch, didn't quite understand the atrocity being performed. If he did, he wouldn't have wanted to continue the arms deal with the Kingdom of Saudi. He didn't want to let the murder of journalist Jamal Khashoggi get in the way of American jobs. He said it was only one man. The conditioning we all receive disrupts better judgment.

"The devil prowls around like a roaring lion, seeking someone to devour."

The Kingdom of Saudi is guilty on 2,977 counts of murder and more than 6,000 counts of bodily injury as a result of the attacks of the World Trade Center on September 11, 2001.

They are also guilty of extortion and ransom of more than 10 trillion dollars in American currency (cash) that was delivered and believed to have disappeared less than a month after the attacks.

Fifteen years later, on September 13, 2019 the key names were released by the Department of Justice in the 9/11 civil lawsuit. All but one name was released to the attorney representing the families of the murder victims on 9/11. I wonder who the blacked out name is? Shouldn't be hard to figure out once we know the others.

There will be events the media will sensationalize to help you forget about the Kingdom of Saudi and a world wide pedophilia cult—stay focused.

I pray these families set aside monitory compensation from the civil lawsuit disclosure and let the world know the names of these murderers so we can destroy these demons. They must be dispatched from our existence.

The Kingdom of Saudi and the Communist Party have infiltrated and are covertly attacking the United States with their psychologically conditioned employees. We've been living in a dictatorship where a president can declare war without congressional approval or even publicly consider mandating anything.

Their soldiers are the businessman and the construction workers along with the boot of oppression from the police to steal private property and resources by force.

Their uniforms are fire retardant clothing and hard hats designated in color by rank. The police are in tax payer issued uniforms, but their pay comes from the oil companies—the Kingdom of Saudi. That makes them mercenaries.

Mercenary—a professional soldier hired to serve in a foreign army, or a person primarily concerned with material reward at the expense of ethics and others.

Mercenaries shall not be treated the same as Peace Officers—no matter what costume they wear. A Peace Officer ensures your constitutional rights. Standing Rock is a good example of these turn coats.

The dehumanization and the militarization of our police force has amplified the compulsive coercive intervention of safety.

Anyone who gives up their freedom for safety deserves neither.

The police murder us and the judges take care of their own, acquittal after acquittal. It's to provide fear and anxiety to produce for the Tax Farm. You are guilty until proven innocent.

Judges are no more better than mercenaries, not royalty, and should be treated as such. For it is above as it is below. Most judges (especially in Texas) pay money for their position of appointment—we must eradicate this problem.

Judges have the power to steal your possessions through the legal system and do all the time—they are criminals themselves, cross dressing. Many judges, lawyers, and their families are a part of a Luciferian society that causes irreparable damage throughout our culture because they can do as they choose without consequence.

These ideologies are handed down from our masters to the corporate psychopaths and are how they manipulate the fearful worker into becoming an accomplice in pollution, theft, violence, and murder. It is willful destruction of one human being by another.

These business men lack the true warrior spirit. They are mere duplicate parasites in suites, separating themselves from you with an e-mail and alcohol and drug induced lunch meetings.

If you are as strong as you believe you are—show your true colors. It will make your painful life more exciting. All will be revealed anyway. You might as well have some fun before your fall. Most of you are lily-white, so I understand your apprehension.

The oil business and eminent domain are really the weaponization of our air, food, and water.

These Saudi owned oil businesses—with direction from the true owners—are paid to strategically frac and destroy our water supply with H2S, Benzene, and a great number of fugitive emissions that occur in crude oil. <u>Only 1% of the water on earth is drinkable.</u>

Fracking - a pressurized force into the earth to fracture rock and force fluids and gases to the path of least resistance. Most of the time that least resistance is our water supply and private wells. They destroy clean water wells for hundreds and thousands of miles with one operation.

All through Texas and up the Mississippi River Valley, on through the Dakotas, the fresh and easy to get to water is poisoned—causing earthquakes even in Texas. It has such a concentration that the water combusts into flames from the amount of fugitive hydrocarbons emitting from peoples kitchen faucets.

Those fugitive emissions are leaking into peoples homes and slowly causing mental difficulties such as dementia (destroying the blood brain barrier) and causing anxiety. These are the very same waters that are suppose to nourishes our crops.

<u>To destroy a water supply is an ancient act of war</u>, and we are indeed under attack by these people—we are viewed as slaves beneath their feet. They pay us to weaken ourselves—mind, body, and soul—while paying time and a half. We are at war with anyone who poisons our air, food, water, or consciousness.

The Kingdom of Saudi and the Communists Party has outwitted America in every way. Now we are becoming more aware of the Kingdom of Saudi's desire to treat the resources of the great State of Texas like a cheap prostitute willing to be defecated upon and beg for more for just a dollar more. That awareness is becoming more vivid by the second.

People fall victim to a prescribed ideology of "pollution smells like money"—more like, "smells like bloody sinus." Anyone who subscribes to this philosophy is your enemy and the enemy of all living beings and is indeed a victim of manufactured consent.

Manufactured consent - the effective and powerful ideological institutions that carry out a system-supportive propaganda function, by reliance on market forces, internalized assumptions, and self-censorship, and without overt coercion by means of the propaganda model of communication.

The Kingdom of Saudi wants to steal our land and the Gulf Coast because water is the next exploited resource.

How would you feel if the Gulf Coast was militarized by private security firms (mercenaries) to protect their assets—the ocean? Not to protect it from being destroyed, but just to protect their claimed property from you.

Dirty water means you pay them to clean it for you—desalination. Desalination will give complete control over life of all living beings. They want you to go through them to have what is necessary for all life. The life source given to us by the Lord, Universe, and Mother Earth. All living beings are the true owners of clean water. As I said before—only 1% of the water on earth is drinkable.

The Palestinians suffered from the same tactics: Befoul their water supply, then spray their homes with raw sewage and bio-hazardous waste, and finally, move them away from the coast. Now they live in an open air prison (reservation)—unable to leave by land, air, or sea. They have no food or clean water, medical supplies are air dropped by sympathetic countries, and they are slowly being squeezed to extinction. Look at a map of Palestine from 1950, then present day - you will then know it is true. <u>If they push us to the center of the United States - we are lost.</u>

The Kingdom of Saudi is a well known—timeless and traditional—abuser of women and children. One of the more well known and recent incidents involved a man who intended to kill his wife after she answered the phone without his permission. He beat her into a bloody, swollen head, and it was seen as tradition—not a criminal act.

The abusive male guardianship system is the indoctrination of their young. This philosophy gives male Saudis an opportunity to learn oppression from birth and there are few places for Saudi women to turn.

But wait, the gift just keeps on giving. They start with women, then move to the rest of society. They want to bring it world wide—world wide repression. Some women try to escape and most of the time, are killed for their troubles.

A Saudi man controls a Saudi woman's life from her birth until her death. All Saudi women have a male guardian. This is normally a father or a husband, but in some cases a brother or even a son has the power to make a range of critical decisions on her behalf. They are treated as permanent legal minors until death.

Saudi Crown Prince Muhammad bin Salman wants you to view him as a women's rights reformer, but nothing could be further from the truth when the authorities—under the control of the Crowned Prince—hunt down fleeing women and torture women's rights activists in prison. Saudi women are indeed slaves. Michael Page, deputy Middle East director at Human Rights Watch, says they are the most draconian in the Middle East.

Draconian - name from a 7th-century B.C. Athenian legislator who created a code of law written in blood; so cruel or harsh—death was prescribed for even minor offenses.

The Saudis restrict the movements of its female population—very Genghis Khan of them. Women cannot apply for a passport or leave the country without their male guardian. The Interior Ministry imposes and enforces these restrictions.

Many resort to hacking into their male guardian's phone to change their travel permission settings or run away from family members while outside the country.

Saudi men may have up to four wives at one time, and have no minimum marriage age. In many cases Saudi men marry 8 year old girls.

These bastards are truly the wild ass among men. His hand is against every man, and every man's hand is against him. He indeed dwells over how to be against all his brethren.

The assassination of Jamal Khashoggi was the final straw.

What I'm seeing from this action is no one can leave the kingdom. The people are the property of the Kingdom of Saudi to do with as they choose to the end of their existence—which is why Saudi owned companies are self-governed like Exxonmobil.

If interrogating someone means you cut them up—then I want to interrogate the Kingdom of Saudi and its king.

No amount of fallatio from Bloomberg will soften this blow. The news outlet—not the stop and frisk guy.

Why do the European, Eastern, and Middle Eastern societies say Mother Earth is a haughty, watery monster in need of taming and revel at her groaning?

She was a lot bigger—leaving behind a portion of her self fragmented in the Hammered Out Bracelet. From the male planets Cronus (Saturn or Satan), to Zeus and the Titans. Even much earlier—the Sumerians (later Babylonians and Assyrians) and their Anunnaki story—all so-called gods raping and murdering Gaia (Mother Earth).

Is Mother Earth our enemy? I find it hard to believe that what nourishes us is also our enemy. A woman is the vehicle of life, the giving of birth and nourishment. They are identical with Mother Earth.

Native American cultures can somehow live in harmony with Mother Earth—why can't the eastern cultures? <u>Because cultures that mistreat their women mistreat Mother Earth (Gaia) as well.</u>

Henry Ford was rewarded for creating a machine that drained Mother Earth—as did Andrew Carnegie. Foreigners come to America to live Andrew Carnegie's American dream; I tell them to look in their own back yard and when you smell something that smells like bloody sinus—think of Ford and Carnegie.

In short, we must annihilate the Kingdom of Saudi's military; imprison or kill their so-called king and royalty; pump their oil back into the ground; and take all their gold, money, and tangible assists (foreign and domestic) as collateral for clean up and damages against the people of Texas, the Americas, and the World.

There ain't but one king in Texas—and that's Bob Wills.

All palaces are built on the ruins of the bowers of kings, and the hunger for absolute power is a natural disease of royalty.

Government by kings is a heathen concept copied by the children of Israel, prior there were none in the world of Yahweh. Abraham came from a scorned priest's family from Sumer. They didn't suffer under the Akkadians that arose after the fall of Sumer.

As Thomas Paine wrote in his book *Common Sense*, "It was the most prosperous invention the Devil ever set on foot for the promotion of idolatry." The Heathens paid divine honors to dead kings and, eventually, the Christian world did the same with the living ones.

Gideon and Samuel disapproved of governments by kings. Three thousand years after Moses, the Jews under national delusion requested a king. It is the pride of kings which throws mankind into confusion, it can't be defended on the authority of the Scriptures according to Gideon and Samuel.

Gideon was victorious against the Midianites, and the Jews cried rule over us, thou and thy son, and thy son's son. Gideon tried to tell them the Lord rules over them. Gideon didn't decline the honor, but denied their right to give it.

The Jews told Samuel to make us a king to judge over us like other nations they were so eager to be like other nations. Samuel tells them a manner of a king that will reign over you will appoint your sons for himself for his chariots and horsemen. And he will appoint your sons over thousands and captains over fifties, and will set them to reap his harvest, and make his instruments of war—like the Military Industrial Complex.

And he will take your daughters and make them cooks and bakers. And he will take the best of your fields, vineyards, and olive yards and give them to his servants—like oligarchs. He will take a tenth of all you have and give it to his officers and servants—like taxes.

Nevertheless, they wanted a king to rule over them, judge them, and fight their battles.

Samuel had to call out to the lord for thunder and rain to shut them up, and they begged not to die because they added to their sins by asking for this evil. Either the Lord protested against monarchical governments or the scriptures are false.

All men are created equal. It's ignorant to allow imposition on prosperity, and perpetual preference of a family whose children are far unworthy to inherit.

If we trace back monarchical families we see their rise as chieftains among plunderers who increase their powers by force, and have the meek and mild purchase their safety with coerced gifts.

The evils of hereditary succession should concern mankind. Men who look at themselves as born to rule, and others to obey soon grow insolent.

For one honest man is of more worth to society, and in the sight of God, than all the crowned aggressors that have ever lived.

It's time to turn on our Arab fathers and reject all so called Babylonian superiority.

Everything the Kingdom of Saudi provides is immediately dangerous to the life and health of all living beings on the planet—acute and chronically.

It looks like they actually work for some other being that wants to destroy the earth by fire, a being that likes its atmosphere very hot and dry.

This being is not the source, an imposter only. This lie comes from the sons of Issac and Ishmael and provides an acceptance of the outcome, the destruction of the earth. I have a deep desire to feast on the flesh of these beings. I'll bet it's a lot like chicken.

The true source is the unmoved mover moving in all the moves.

The Kingdom of Saudi are the lowest vibrational beings of density on the plant and anchor human progression.

I'm sure they will try to hire mercenaries, another nation, or continue to use the United States government to put up a fight for their interests. MENE, MENE, TEKEL, UPHARSIN. In the end—the kingdom must boldly meet their reversal.

I pray to the lord and the powers of the universe to choose me to pierce the black heart of these demons. I'd be willing to negotiate a withdrawal with the outcome listed above and maybe they can keep their life.

US soldiers are already in Saudi Arabia. We already have the logistics on our side.

It will be the beginning of the end of the vampiric draining of Mother Earth's blood-source. It will also be the greatest transfer of wealth since the church abandoned Constantine.

For some of you less intelligent humans reading—it is obvious that there are good people in Saudi (much like Jamal Khashoggi) and other Arab countries.

When I say the Kingdom of Saudi—I'm speaking of those who have the money, power, and influence to steal, rape, poison, and murder without consequence.

We must focus on eliminating our weakest link in human progression before we can redirect to truly benefit all living beings.

The time is now.

Water

XV

Water is the life giving substance that all living beings require for survival. Without it most living beings would last only a few days. We take water for granted because we can get it from our kitchen sink or the many gas stations at every corner.

Our planet and our bodies are made of mostly water. Only 1% of the world's water is suitable for drinking. It took millennium to filter and fill aquifers —and it could take just as long to replenish them.

In North America the extraction and consumption of water is greater than the supply. Most excessive water usage is from industry and agriculture— the other is wasted because of leaking water mains.

We used to have artesian springs all over Southeast Texas—until the paper mill in Evadale started using an excessive amount of water.

Cape Town, South Africa was the first major city to have a countdown to day zero (day zero is the day of no available water supplied)—4 million humans without water and a great number of other living beings without as well.

It led to water rationing and long lines at city water stations. You thought it was scary when toilet paper was hard to get? Don't show up thirsty.

More major cities are counting down as well—Mexico City, Tokyo, Barcelona, Bangladesh, Istanbul, Beijing, London, Jakarta, Melbourne, and Sáo Paulo to name a few.

North India reduced its water supply by 29 trillion gallons in just a decade. In February of 2019 a so called terrorist attack in Kashmir provoked New Delhi, India to criminally retaliate by cutting off some river water supply to Pakistan. Any excuse will do. The dam on the Ravi River allocates water to Pakistan by treaty brokered by the World Bank—a privately owned bank.

The one and only treaty in the world that has ever held up was between the people of Fredericksburg, Texas and the Comanche. Every other treaty that has ever existed has been broken with money, greed, and murder.

Treaty should be defined as—a signature for your later theft and death.

India is the source of a large portion of our troubles on earth. Consumption and waste is the Indian way. They hide behind a facade of tolerance and peace—when, in reality, the water India withholds belongs to all living beings down stream as well. They are over populated and trashing our ocean with petroleum products. We need to get a handle on India before they destroy everything.

Yemen and Syria are out of clean drinking water. More than 14 million people—mostly Yemenis—are living in famine conditions. It is clear the siege led by the Kingdom of Saudi and the United Arab Emirates is a successful attack to obtain all coastlines, international shipping lanes, such as the Ras Isa oil export terminal, the Sues Canal, and the fertile lands of the Tihama plain in Yemen.

The Kingdom of Saudi and the UAE call this attack the "Golden Victory"—smells like money.

Aerial bombing and ground attacks have damaged the clean water and sanitation infrastructure which has caused a Cholera outbreak.

They have to truck in water for survival or leave the area if they have a place of refuge. Fires in California and hurricanes on the Gulf Coast hardly equal the pain these living beings suffer.

We can leave for the fire or hurricane and come back with supplies. Most do not have the means for travel or supplies and are not permitted to enter other countries. Both sides aim their destruction at humanitarian supplies—a kind of attrition.

We are looking at a humanitarian collapse. Our little chicken-shit hurricanes and power outages are no match to the pain these people suffer because there is an end to our blight.

Attrition - the action or process of gradually reducing the strength or effectiveness of someone or something through sustained attack or pressure. Sorrow, but not contrition, <u>for sin</u>.

The Kingdom of Saudi and the UAE have militarized these coastal countries for access to water. Syria and Yemen have to go through the Kingdom of Saudi and the UAE to receive what is necessary for life and travel—the life source given to us by the Lord, Universe, and Mother Earth. All living beings are the true owners of water. As I said before—only 1% of the water on earth is drinkable.

Even Goldman Sachs knows water is the new oil exploitation. Hedge funds are buying up water rights globally. Do these suits have the ability to control the actual tangible asset? Can their claim be asserted?

Santa Ana tried to claim half of North America up into Canada just by saying it was his. He couldn't defend it or physically travel to it. He was just calling shotgun so he could sit in the front seat and maybe he could have stayed sitting. That is if someone stronger doesn't take it from him—and they did.

Who ever thinks they own your right to water is the enemy of all living beings and must physically be punished.

And now the Syrians and the Yemenis suffered the same tactics Israel executed on the Palestinians. Befoul their water supply, spray their homes with raw sewage and bio-hazardous waste, and move them away from the coast.

Now they live in an open air prison (reservation), unable to leave by land, air, or sea. No food or clean water—medical supplies air dropped by sympathetic countries. Slowly being squeezed to extinction.

Look at a map of Syria to the North and Yemen to the South—you will then know it is true. <u>If they push us to the center of the United States—we are lost.</u>

The Kingdom of Saudi—with help from Israel—owns many coastlines including the southern Mediterranean Sea and the clean water supplies. They know water is the next oil exploitation, and has always been the plan to obtain. Next will be Iran and Turkey—the Kingdom of Saudi started locally and wants to go globally with Babylonian superiority.

Why is there such a reduced amount of clean drinking water in Africa, the Middle East, and Mongolia?

Think of it this way: If you take a quart of oil out of your car—what's going to happen? It will over heat. If you replace it with salt water what's going to happen? Friction from pistons will heat, warp, and cease. How about taking two or three pints of blood from your body? How well will you function?

<u>Oil is the coolant system and blood source of planet Earth—a living being.</u> The more we remove the coolant system—the closer we collapse toward the earths hot core—the hotter it gets.

The removal of the coolant system (oil) will allow the core of the earth to cook and evaporate what water and moisture we have. It will evaporate all clean drinking water, destroy all vegetation, and kill all life on earth.

Without lubrication the Tectonic Plates will cause more massive earthquakes and volcanic eruption of biblical proportion. We must keep the core cool with oil for survival of all.

What being could convince us to systematically remove our coolant system for ultimate destruction? All for fiat money? Most planets in our solar system have salt gathered in condensed places. This comes from the removal of brine water leaving only salt. What is this parasite trying to make our constellation uninhabitable? Are we next or are we the last? I'm ready to put these beings on the menu.

<u>Continuing to remove oil will cause the earth to be destroyed by fire from the core.</u>

There are thousands of oil rigs in the Gulf of Mexico operating with few, if any at all, guidelines on emitting pollution. Mostly because of the vastness, there isn't anyone out there to complain of damages and resources to enforce are limited. Most of the oil rigs in the Gulf of Mexico are owned by foreign entities (Saudi, Russia, Dutch, Chinese).

The removal of our coolant system (oil) in the Gulf of Mexico has caused the gulf floor to heat up from the core, causing hurricanes to intensify in already shallow water. Hurricanes are intensifying on the African coast because of the removal of the coolant system—with the land becoming more desolate and bereft. The more desolate and bereft—the bigger the dust storms that blanket North America will become.

We will then see more respiratory infections, and our cells will excrete more exosomes. The more exosomes we excrete, the more our leadership will try to control us for so-call safety. Covid is pollution from the continent of Africa and from Mexico.

<u>Anyone who gives up their freedom for safety deserves neither.</u>

On June 1, 2019, PBS News Hour reported the United States experienced 500 tornadoes in just 30 days—some of which killed people in Alto, Texas at the Caddo Mounds on a bright and sunny day. It ripped through the pines in a flash.

Not far in Bryan, Texas, 5 more people were killed by mysterious tornados. A mother and two children were killed in Erath County—swept away in a flash flood on April 24, 2019. In an instant, tornadoes ravaged the area like it was designed as an attack.

Are there beings geo-engineering our planet to cause harsh weather conditions as a weaponization of weather to push us from the coast? Or are the polluters terra-forming for beings that like it very hot and dry and can only breath ground level background ozone (O^3)? Are Californians being burned out of the state because of overpopulation, or is it a massive land grab for a rail system, or is water the new oil exploitation?

Most all fires in California and Australia are in proposed area for the new rail system. It appears they don't want to pay the property owners full price. I feel for these humans, but their conditioning to communism has allowed this to be so. They want to be a good people and know not what that is. They can think not for themselves and are prescribed an indoctrination. The only thing that can help is the belief in a higher power and the Constitution of the United States, freedom in both.

Some how the fires stay on the U.S. side of the Canadian border. There is documentation of laser beams coming from Canada to the point of ignitions of wildfires. If this is fully proven, Canada will be the next U.S. frontier.

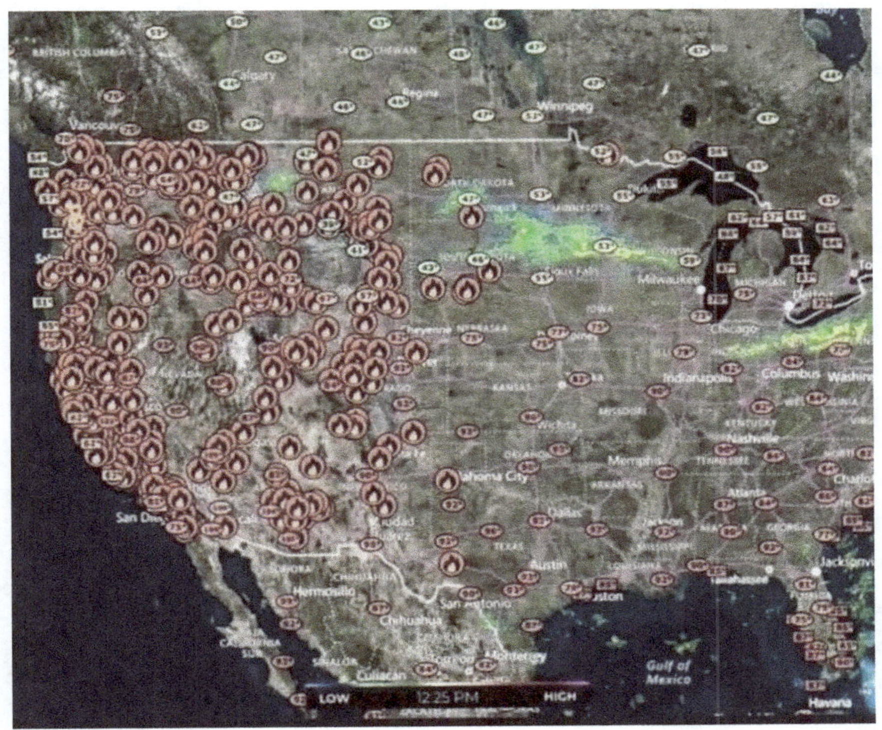

California is out of water. Existing law requires the state to achieve a 20% reduction in urban per capita water use in California by December 31, 2020. This coming from the California Legislature - AB 1667 Water Management Planning.

I guess smoking and burning Californians out of homes and the state for a high-speed rail system without having to purchase expensive land is one way to reduce the population if you do not have the intellect to stop construction of inhabitable structures. Taxing them has worked as well. Texas will start the out of state tax with a back date.

The Giant Sequoia trees in California drink 100,000 gallons of water a day—one tree drinks 800 gallons. Trees absorb the carbon molecule from CO2 (carbon dioxide) and release the O2 (oxygen) back into the air. Scientists can now see this tree is shutting its pores to help conserve its water usage. The effects from shutting its pores are not being able to absorb carbon dioxide. The Sequoias, among many trees, are losing natural function—the very essence of their function—because they're dying from dehydration.

The blight on the oaks in Austin, Texas is a clear indication of the severe reduction of water. To say the word " blight " in this instance will be a verb - have a severely detrimental effect on. The leadership in Austin will have you believe it's a noun—a plant disease, nothing could be further from the truth.

The cause of this blight is from overpopulation draining this small water supply. There are too many people in Austin drinking, showering, and flushing toilets. We know this to be true because Austin doesn't have enough water pressure to put fires out in East Austin. The consumption is greater than the supply.

If you're really from Texas you know the rule of thumb when it comes to cows per acre. Where I come from (the bayou) we have plenty of water for our population. However we can only have two cows per acre due to grass/water consumption and waste output (cow shit and piss).

With less water for dilution the waste is more concentrated in dry areas, making chances of Cholera exponentially grow.

Austin, Texas is only a few steps from becoming a third world country like Syria or Yemen. There are more than 2 million people on a tiny water supply flushing toilets, taking showers, and drinking water—everyone will do at least one of these functions. They are also polluting other cities on this water shed like San Antonio.

San Antonio is fully dependent on the Edwards Aquifer for municipal and daily use. Similar to Pakistan—their troubles derive from being down stream from Austin's trash.

If I were San Antonio, I'd sue Austin for damages of person, property, and the destruction of livelihood, for starters. They are physically being poisoned and their property will drop in value from the lack and quality of water. If that didn't work, I'd send a large group of people to aggressively and physically halt construction and arrest the liberties of the leadership in Austin.

Midland—Odessa pollutes the Edwards aquifer as well. They have removed most of their coolant system with fracking techniques that destroy the water supply to San Antonio. Most of the oil and land in Midland—Odessa belongs to Chinese companies like Surge Energy and Yantai Xinchao to name a few. So where is the value other than a few people being paid an obscene amount of money by the Communist Party to poison ourselves?

Communist China is also the largest land holder in America. How's that for Texas pride? As a matter of fact—most billionaires became billionaires because of their connection with the Communist Party. They were born in America, they are white, and all their businesses have Chinese symbols as logos on their equipment. The top tier of the communist party are all billionaires—the most billionaires per capita. Everyone else has to suffer the pain of really being controlled by the Communists. We will revisit this.

Dripping Springs, Texas—outside of Austin— ran out of well water. Residents' private wells went dry because Austin drank more than their share. They had to truck in water for daily use. Central Texas has too many cows (consumers) per acre. Central Texas can't make it with droughts and heavy consumption.

Now Austin wants to put in a private toll road (MoPac South) for access to more habitable construction. We have 10 lbs of shit in a 5 lbs sack. If we let them—they will destroy it all.

Wall Street investors like BlackRock own 1 in 7 homes in America, raising prices on homes across the country. Foreign buyers spent $634 million in Austin, Texas real-estate since the pandemic with the medium price of $428,000. China and Mexico made up 14% each, Canada 6%, India 6%, and Armenia 5%. We will give more and greater remuneration to a home-owner with homestead exemptions, meaning they live there. Investment homes will float the tax base, especially corporate owned for investment.

The river and water authorities will report what they are told to report. We don't need statistics to see its drying up—use your eyes and brain.

We must reduce this population immediately by removing a few large and undesirable companies from this area. We could start by removing four large undesirable companies (such as Google) for every one desirable.

What is desirable? Both parties to any exchange benefit from birth to death? Products that poison not our food, water, air, or consciousness from the womb to the tomb? One that intervenors gain not at the expense of subjects who lose in utilities? One that provides harmony to the market? One that prevents the scramble to be a net gainer rather than a net loser or prevents the scramble to be a part of the invading team instead of the victim? Vehicles that can be maintained by the purchaser. Entertainment is a weak

market, so are tulips. The tech world is mostly entertainment.

We must stop construction of sub-divisions, apartments, condos, hotels, and any type of condensed housing unit all through Texas. We must take the business license from any company who requires proof of vaccination. We must cease all assets of any company partnering with the communists.

Texas must adopt new regulation to severely limit housing on water supplies. Dryer areas on and west of the Balcones Fault Line, Houston, and south of Houston must be limited to 8 acre tracks (per household) minimum. Places that have water must limit construction to at least 2 acres per division. Everyone else can live somewhere else. We must leave room for natural born citizens to multiply and grow.

We will not take people's property already liv-ing smaller—we will stop dividing land up smaller than said limitations. If there is a run to get land de-veloped before deadlines, the only thing that will help you will be a time machine because the line was drawn twenty years ago.

If you live in Beaumont, Texas, you will see gas stations on top of gas stations at every corner and more to come. Anywhere Beaumont has something beautiful, a fucking gas station convenience store is in its view. We have the historic Mildred building—across the street is a fucking gas station with homeless people kicked back. We have a beautiful Event Centre—looking down the Great Lawn is a fucking gas station convenience store with homeless residents.

They developed beautiful homes out west of Beaumont on highway 90 and added a fucking gas station convenience store at the gate because they are white trash. The gas stations are usually named "Kris Kross" or "Krooze-in" and serve fried chimichangas or egg rolls. The smell permeates two hundred feet and with the lords favor— your credit card will work at the gas pump. If not, you'll have to go in to pay and smell like a fried chimichanga for the rest of your life.

We have strip mall with smoke shops, pay check cash stores, and unlicensed massage establishments. Beaumont also has apartment complex on top of apartment complex springing up—and now we have real traffic jams in this one horse town. Beaumont, Texas has not the infrastructure to support these people who care not for our comunity nor can they construct road to match the volume of people already living here. We simply have not the room for beehive living.

This disgrace comes from the City of Beaumont, development boards, and planning commissions that consist of white trash appointed by our former mayor and city council. Having money exempts you not from being white trash.

All said entities consist of mostly bankers and realtors who believe that quantity is better than quality—because they are white trash. These people stand to make a lot of money from the plans developed by themselves. Their greed is exceeding and they care not for our community—they only care about their own bank accounts. Maybe we can develop gas stations and apartment complexes next to their houses? We can look down into their homes from the apartments—we can see how they live.

The slum lords that own the building caresses downtown Beaumont prevent downtown living in these beautiful building already constructed. Why buy a building to watch it crumble? Because they are white trash. Maybe its because downtown is neighboring Exxonmobil that is in its final stage and is ready to blow up at any moment. Old families that own the majority of properties are claimed to be responsible for the building caucuses around town. Some places need not to be inhabited because of the volume of people in a designated area—maybe there needs not to be a building there in general.

Texas and Texans have not the ability to understand beauty, only money—white trash. This is happening all over Texas and Louisiana. Do you know why you love New Orleans? Because it's old and beautiful, and it's acceptable to be belligerent.

Watch a few episodes of "Hoarders" and you will see the depressive want for clutter caused by the feeling of loss from trauma. They can survive not psychologically without clutter. It reminds me of the 1986 movie "Labyrinth" when Sarah comes in contact with the Junk Lady and finally realizes it's all junk. Most hoarder believe they can save an outdated World Book with information that isn't true anymore, and when everything goes to shit, they will have the information that will save us—mental Illness.

Austin doesn't have the resources to remove the Zebra Mussel taste and smell from the water. What else can Austin not remove from our drinking water?

The run-off in Lady Bird Lake has killed many animals—wild animals, livestock, and pets—a real red tide. My own dog got very sick from swimming where Barton Springs Pool flows into Lady Bird Lake—on the banks of Zilker Park.

Ignorant local authorities claim the Blue Algae is what's killing animals. The Blue Algae is a mere indication of chronic water pollution. It's like saying your headache is what made you sick instead of a symptom of being sick. Urban development and agricultural runoff is exponential on the Colorado and Rio Grande river. There are more than 2 million tons of sewage, agriculture runoff, industrial waste, and plastic going into the rivers and ocean everyday.

There are parts of the Rio Grande where you are strongly advised to not submerge your head or enter the water due to severe pollution and agricultural run-off. It is dangerous to life and health.

The Colorado river rises in Dawson County—on the Texas/New Mexico boarder—and flows over 600 miles to the mouth of the Matagorda Bay. It fills many lakes including Buchanan, Inks, LBJ, Marble Falls, Travis, Austin, and Lady Bird Lake. It is a wonderful natural life giving resource for all living beings—animate and inanimate—to enjoy.

However, arid drylands farming, the over populating of Central and West Texas, and the removal of our coolant system (oil) with fracking techniques has a detrimental effect and will severely damage and disrupt all life in this region for years to come.

Growing alfalfa in any desert is bullshit. There isn't an old farmer in overalls hoeing rows—it's a conglomerate of psychopaths. These are $100 billion dollar foreign companies who find it funny how they can buy and throw away America as they choose. We must end major agriculture in arid, dry places.

There's a lot of talk of building a wall to keep illegal immigrants from entering our country. As long as we are under a Capitalist system we must keep an illegal influx of labor from diluting our wages and freedom. Building a wall would mean someone will be cut off from the Rio Grande.

The Lord, the Universe, and Mother Earth provided the river water for all living beings in the vicinity. Who will be denied access from what the source provided for all? It's a horrible idea and anyone who participates in the separation of this life source will find their life will be damaged in mysterious ways. There will be other ways of correcting this problem.

Our leadership in Texas should be ashamed. I'm sure they are not because money and greed is their god. They may claim to be freedom fighters and have a big red "R" for Republican in front of their name—in reality they replace god and the good for all

with money and appear to have communist characteristics.

Publican - ancient Roman collector of taxes,
and now we have Re - publicans.

Re - once more or with return to previous state.
Look in the Bible, Matthew 5:46 for better understanding.

Communist China has a strangle hold on Texas
and its leadership. You can witness these meetings and
photo ops taken with the Consul General Li Qiangmin and Governor Greg Abbott and 10 State Representatives who met separately on February 6, 2017.
Who are the state representatives?

"They shared views on cooperation between
China and Texas in the fields of energy, agriculture,
and medical care, and cooperation between Chinese
cities and Austin on creativity." says the Ministry of
Foreign Affairs of the People's Republic of China.

Austin Mayor Steve Adler (the privileged Covid
hypocrite that breaks his own rules) "discussed the
new era of increased direct city-to-city interaction and
cooperation" with the communist party on December
18, 2019 just two months before the pandemic.

How is it that Communist China can indoctrinate Texans at the Flying Tiger Academy in Houston,
Texas? Communist China has imprisoned the largest
population on the planet and now they have come to
Texas to expand their tentacles. The Chinese people
are murdered and imprisoned just for speaking

against the Communist Party—and they have free reign in Texas?

The Communists are the very opposite to freedom and liberty—the very essence of what we in the United States are made from. Fuck the Communists and their Texan accomplices.

We must run the Governor of Texas all the way out of Texas. He understands half of person and property, mainly money—who is owes what to whom. The Governor received his education on the subject from his lawsuit. He lacks the ability to understand true freedom and where my rights end your rights begin. Greg Abbott has aligned himself with the communist demon energy.

He has allowed many companies rights to overlap and supersede my rights by suing the federal government to allow our water and air to become what it is today—toxic. Here are a few examples:

Aug. 23, 2010 - Texas and North Dakota sued the EPA over rules that limited how much sulfur dioxide could be in the air. We have 78% nitrogen, 20% oxygen, and 2% other. The other consists of dust, pollen, and "other" naturally occurring elements. If you can't operate within that 2% along with the other natural occurring elements—then you are providing damages to all living beings.

Sulfur dioxide is causing respiratory infections during the winter time. The media will try to convince you that you're dying from Covid, but the true cause will be sulfur dioxide.

Status: Loss   Cost:$33,718

Sept. 11, 2013 - Texas and three other states sued the EPA for taking too long to decide which areas of the country had too much sulfur dioxide in the air. The state also intervened in lawsuits filed by the Sierra Club on the same issue, and the cost reported includes figures provided by the agency for all such lawsuits.

Sulfur dioxide is being released by dilapidated equipment in petroleum refineries, cement manufacturing, paper pulp manufacturing, metal smelting, and coal power plants. It damages billions of dollars in crops and damages waterways making it unsuitable for fish to survive—we need fish for our own survival. All said industries can operate without poisoning us—it just cuts into their profit. It costs too much because the market is a lie—the only way to profit in this market is to steal and damage.

It is the cause of what people believe is Covid.

Status: Pending   Cost:$29,163

July 19, 2012 - Texas sued the EPA for adding Wise County, a largely rural county, to a previously defined region of North Texas counties whose smog pollution levels violates federal law.

The population of Wise County or any rural area has nothing to do with the amount and concentration of poisonous air traveling for hundreds of miles depending on the wind patterns of summer and winter. Being just a few miles from Dallas/Fort Worth, which Wise County does reside northwest of

Fort Worth, is a clear indication the residents of Wise County are being poisoned. I will also bet that learning disabilities, drug addiction, and chronic illness is rampant in Wise County. It isn't Covid—its pollution.

Status: Loss   Cost:$25,122

Dec. 22, 2015 - Texas filed suit over the EPA's tightened standards on ground-level ozone—which aims to crack down on pollution coming from factories, power plants and vehicle tailpipes. Texas Attorney General (Tangle Eyed) Ken Paxton said the rule isn't supported by scientific data and will unnecessarily harm the state's economy.

We are dying from a massive amount of ground-level background ozone coming from Mexico in addition to our own. Use your eyes and brain to see that smog is killing everything. My right to clean air supersedes any economy. A minute or two is all we can go without air, economies come and go. I'm sure Tangle Eye doesn't believe that because money is his god, and he along with others will convince you to keep looking at Covid instead of the root cause, which is pollution.

Status: Pending   Cost:$20,850

Oct. 15, 2012 - Texas challenged the EPA's new restrictions on pollution from hydraulically fractured natural gas wells. They are destroying our water supply and it's an ancient act of war.

Status: Withdrawn   Cost: $5,448

Jan. 17, 2017 - Texas and 13 other states sued the federal government to block a federal rule limiting coal mining near waterways. Tangle Eyed Paxton said "the federal agency adopted the revised rule without the participation of the states."

Sulfur dioxide dissolves easily in water to form sulfuric acid, sulfuric acid is a major component to acid rain. Acid rain damages forest, crops, changes the acidity of the soil, and makes lakes and streams unsuitable for aquatic life. Killing fish and being unable to grow food is taking away our livelihood and safety net for survival of all. Acid rain doesn't burn your skin, it lowers your ph balance causing cancer and other long term illnesses.

Status: Pending   Cost: Not Available

Jan. 18, 2017 - Tangle Eyed Paxton has asked for a review of a regulation aimed at curbing hazy conditions in national parks and wilderness areas in Texas and surrounding states. The regulation requires states to craft plans about how to go about doing so. Tangle Eyed Paxton said the regulation gives federal land managers powers they don't have under federal clean air laws.

I agree with his concerns of land managers' powers of authority—however Texas has allowed the Kingdom of Saudi and any corporation who will benefit him and his to operate with impunity and continue to beg for scraps. Tangle Eyed Paxton wants there to be no reprieve from the poison. There is indeed nowhere to run from toxic air in North America or the world for that matter.

Status: Pending   Cost: Not Available

Lawsuit information came from the January 2017 Texas Tribune.

Both the Crippled Communist and Tangle Eyed Paxton are slaves or an accomplice to greed—one can't stand up for Texas and both don't want to. They want to help the Kingdom of Saudi and corporate Communist China destroy the working people of Texas. Either way they will be removed from office. The Governor of Texas is half RINO half Communist and has traded his humanity and his connection with the lord for his new religion—Law.

The judicial system has their own language, their own hierarchy, and their own gods. They have completely abandoned the human race. You can tell because they only have the capacity to understand money and property, which makes them ill equipped to have a grasp of anything else, such as deadly air and water in Austin, Houston, Dallas, and El Paso.

The Governor of Texas is also a hypocrite even when it comes to the law—privilege will do that. A tree limb fell on him while he was running through River Oaks in Houston and crippled him. He sued the home owner and tree care company and won the settlement, which shows tax-free annuities in graduated payments.

The monthly payments, which began at $5,000 in November 1986, increase at 4 percent per year, compounded annually. They are currently about $14,000 a month. The agreement could be worth about $9 million, and Abbott is paid by this insurance settlement $150,000 a year until 2022. Abbott helped champion legislation around 2003 that imposed limits on lawsuit awards. Under the recent law, punitive damages—meant to punish gross negligence or bad faith—have been capped at $750,000. He's pretty typical for a Texas cull—even the trees want him gone.

This information comes from the Dallas Morning News.

Texas does have the biggest and best—biggest and best criminals and scum. Why must we suffer Texas scum like George Bush, Rick Perry, and Greg Abbott—who all took more freedom and liberties from citizens than most Democrats?

Because people in Texas believe they are wild and free Republicans—in reality Texans are incapable of understanding the qualifications of being free or a Republican. Low I.Q.s in Texas require control—and Texans indeed have low I.Q.s and are controlled. Just steers and queers around here—you will understand why later in the book.

Former governor of Texas, Rick Perry, is the reason Texas is over-populated by Californians and recruited Space X, which moved its launch site to Boca Chica, Texas. Rick Perry said he doesn't care if the residents of Boca Chica, Texas like progression.

He obviously doesn't care if they breath the exhaust from Elon Musk exploding rockets over their town, because Rick Perry only cares about Rick Perry. Rick Perry has been forcing Texans to receive actions and ideas Texans don't want from the beginning and is a disgrace to Texas. Rick Perry forced inoculate on Texas school children with the HPV vaccine using his families pharmaceutical company. Rick Perry is a communist and must be punished for his vaccine mandate. Rick Perry will now live and die in exhaust and fumes, Rick Perry will be mandated to receive every vaccine and booster mandated.

<u>To destroy a water supply is an ancient act of war.</u>

Atrazine is the second most used herbicide next to Roundup and was banned in the EU in 2004. Atrazine is a potent endocrine disruptor that causes immunosuppression, hermaphroditism, and a complete sex reversal in male frogs at concentrations as low as 2.5 parts per billion—well below the safety claims of the EPAs 3.0 ppb.

Immunosuppression - the partial or complete suppression of the immune response system.

Sound like some things we are going through today? Could Atrazine chemically reform humans to reduce our population or turn the strong into compliant?

Hermaphroditism (DSD or Disorder of Sex Development - INTERSEX) - the condition of having both male and female reproductive organs—androgynous form.

Aromatase deficiency - Aromatase is an enzyme that normally converts male hormones to female hormones. Too much aromatase activity can lead to excess estrogen (female hormone); too little to 46, XX intersex. At puberty, these XX children, who had been raised as girls, may begin to take on male characteristics.

46, XX INTERSEX - the person has the chromosomes of a woman, but their outer genitalia appears male—most often caused by a female fetus being exposed to the male hormones Testosterone—ovarian tumors produce Testosterone.

46, XY INTERSEX - the person has the chromosomes of a man, but the external genitals are incompletely formed, ambiguous, or clearly female.

Congenital adrenal hyperplasia - a group of conditions limiting hormone production in the adrenal glands.

The reproductive organs and adrenal glands make the same hormones. The adrenal glands (located at the top of the kidney) do not function properly because of mutations of the gene for encoding adrenal steroid 21-hydroxylase, which is an enzyme. The mutation comes from Atrazine and other chemicals humans are being inoculated with.

Without this enzyme, the adrenal glands may produce too little cortisol and/or aldosterone and too much androgen.

The owners of Atrazine and Roundup will try to convince you that CAH is a type of inherited disorder called "autosomal recessive"—meaning a disorder that can be passed from the parents to their children.

The fact is everyone in the region of high mutations are receiving poisonous water and air. All generations living and inhabiting together are drinking and breathing the same air and water.

It's a tricky little way to convince you that it's your fault. Your family is less than and there's nothing you can do about it—so shut your mouth and just live with it. Companies like Syngenta, and Monsantos—Monsantos has become Bayer—are the enemy of all living beings.

Why is there Glyphosate in beer and wine? Its the main ingredient in Roundup and 300 million pounds is used in the United States as an herbicide. Roundup is the second generation of Agent Orange. Why is there so much poison in our food, air, and water? If RoundUp is so dangerous, why is it still on the shelves?

Tap water is acceptably poisoned with S7 poisons used to exterminate rats and insects. Hydrofluosilicic Acid is a waste product from the Aluminum and Phosphate fertilizer industries. Concentrated solutions are used in water fluoridation. It is a neurotoxin that effects the nervous system and causes

Alzheimers. Fluoride, aluminum, and oxygen together can pass the blood brain barrier, forming what is called Alumina Compound.

Alumina Compound is the basic ingredient in both Sarin Nerve Gas and Prozac. It causes Thyroid problems, low I.Q., brain damage in babies, various forms of cancer, and neurological damage. Is this another way of occupying human consciousness?

It makes great employees—brain dead subservient drones who can't articulate damages against their person or property. The electrification of 5G with the metallic substances in humans brains can allow technology to send and receive information and thoughts, some could be detrimental.

In 2014-2015, Flint, Michigan was contaminated with lead and a related outbreak of Legionnaire's disease. I wonder if Flint has an Exxonmobil cooling tower around that contributed to the release of this Legionnaire's disease?

Exxonmobil claims it introduces a Legionnaire's disease concoction (which has a skull and cross bones on it indicating deadly) to eliminate said disease from their cooling tower, but are releasing it into our breathing air. The people who enabled the contamination plead no contest to misdemeanor charges and will leave without a criminal record.

Most all of the Northwest is radiated from over 1000 nuclear weapons test blasts—not counting nuclear waste stored in Washington State from nuclear power plants.

Nuclear power plants are over designed steam engines—designed to boil water to create steam to turn a fucking squirrel cage. This is some caveman shit with an appearance of futuristic finalization. The waste far out weighs the benefit. One large earthquake in Washington State's high desert will kill many for years to come. Death will come from the nuclear waste they store there.

Somerville, Texas (Burleson County) residents are contracting various types of cancers—mostly stomach cancer—and birth defects from a railroad tie plant at a rate as much as 40 to 60 times the national average.

The current railroad tie plant owner Koppers Inc., a Pittsburgh-based chemical manufacturer and longtime previous owner Fort Worth-based BNSF Railway (formerly the Chicago-based Atchison, Topeka, and Santa Fe Railway), has operated with malicious intent for 100 plus years and have a complete disregard for all living beings. They are guilty of crimes against humanity and must be severely punished.

This railway company for years had disposed of wastewater in unlined pits, contaminating the Carrizo-Wilcox aquifer and its Simsboro formation. The land beneath the plant remains listed as a federal hazardous waste site under the Resource Conservation and Recovery Act.

The railroad tie company Koppers continues to use various heavy-duty pesticides and wood preservatives, including coal-tar creosote, a tarry, chemical stew. It is a carcinogen well known by the National Toxicology Program of the U.S. Department of Health & Human Services.

Koppers is a publicly traded Fortune 500 company that specializes in manufacturing carbon chemicals from coal tar. It is a good idea to divest in this company to try and salvage your profits. We will be free of creosote products soon.

We have the ability to compress trash into rail ties. Not only railroad ties but telephone and electrical poles. Creosote ties and poles are poisoning our water supply—leaching out slowly. This product lacks benefit to anyone but the manufacturing owners. I feel no pity for Koppers or anyone like them. Shut down all creosote products—failing to do so will make authorities an accomplice. We already have the evidence—we will be watching.

All Railroad Commissions have entirely too much power. It must be stripped from their hands.

We are now in a time where most all humans are paranoid of water contamination—with excessive and compulsive filtering if you have the means. Some try to collect rain water in hopes of gathering a more pure source.

Excessive governments have outlawed the collection of so-called "Illicit rain water" or possessing the ability to produce your own energy. The enforcers and writers of this law are the enemy of all living beings and are guilty of crimes against humanity.

Public Law 105-85 allows chemical and biological weapons testing on civilians without explicit consent.

NASA claims injecting vapor tracers in the upper atmosphere have greatly aided our understanding of our planet's near-space environment. NASA says Tri-methyl aluminum (TMA), Lithium, and Barium make visible the naturally occurring flows of ionized and neutral particles either by luminescing at distinct wavelengths in the visible and infrared part of the spectrum or by scattering sunlight.

NASA has a blank check from the taxpayers of America and have very little oversight. Ex-lawyers, now senators, and congressmen are the gate keepers. A lawyer can't do anything for themselves much less understand NASA, they only have the capability to be lawyers and to project confidence—it's a sad and pathetic affair.

For if these ex-lawyers understood that Tri-methyl aluminum (TMA), Lithium, and Barium cause severe neurological damage such as Alzheimers and low I.Q. - they would sue NASA (a private entity con-

trolled by government). If they do understand then they are accomplices and their silence is consent.

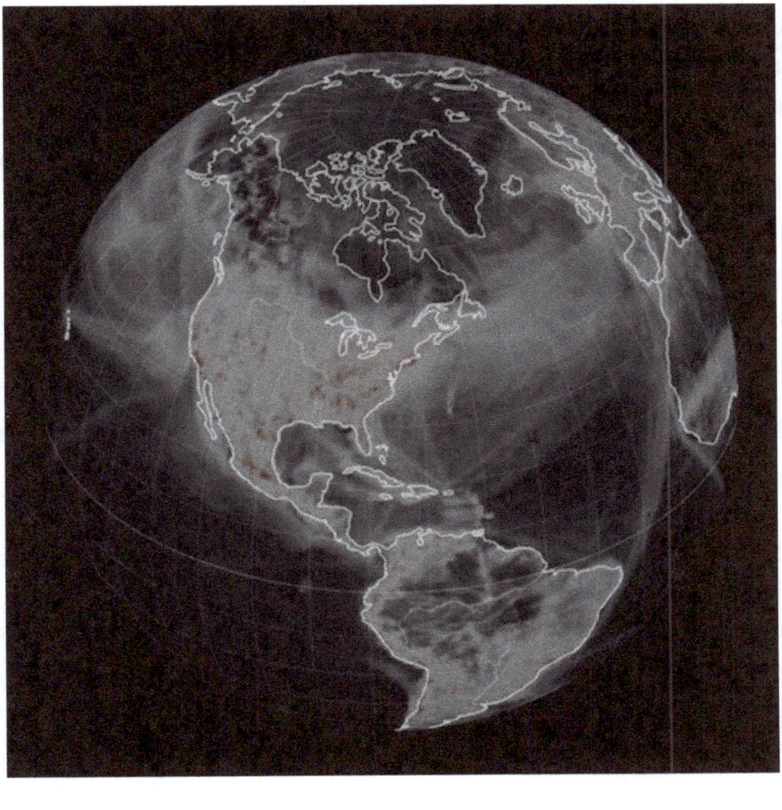

Is this yet another way of demociding humans disguised as science? In hair analysis tests all test subjects show the Barium levels to be off the chart. You can witness this data in the Toxic Elements section of the test, it can passed through the Blood Brain Barrier.

Democide - the intentional killing of an unarmed or disarmed person by government agents acting in their authoritative capacity and pursuant to government policy or high command.

We know very little of what happens in space or at NASA. The average person lacks the ability to reach this threshold.

The Apollo 13 Lunar Module is made of about 5 layers of tin foil—claimed the astronauts involved. Buzz Aldrin explained on the Conan O'Brien Show that the moon landing people watched on television was animation. So what is the vacuum space agencies refer to? Any seamstress can make a space suit? They honored black ladies form the 1950s that made the suits on CBS Sunday Morning program—they were seamstresses. Is there no air? How much air can they carry for an extended stay?

Vacuum - Latin for empty, unoccupied. A space entirely devoid of matter. Vastness, a state of isolation from outside influence.

The idea for a space from which air has been removed or partially removed is a new idea. To displace air by suction of matter into a container by displacement.

When Ross Perot said there will be a vacuum, he didn't mean a sucking sound, he meant an emptiness.

I've never been to space. It appears to be vast, and sound has nothing to bounce from because of vastness. Is there no friction? How did astronaut John Glenn see sparks while riding in the Friendship 7 capsule in 1962? He was only traveling 17,000 miles per hour. Was there no friction, or was it just reduced? Astronaut Scott Carpenter saw the same sparks while in the Aurora 7 space capsule.

We went from Daredevils riding in tin cans to a trillion dollar industry funded with dump trucks full of flaming tax dollars. All for the excitement of the unknown. Pretty much like the first made science-fiction movie in 1902 "A Trip to the Moon" where the intellectuals gather to discuss, then form a plan, then shoot a capsule out of a cannon to the moon, then have an adventure, and then a splash landing—just like we did when we started. It's circus and bread.

It is important to know that the founding member of NASA (formerly Jet Propulsion Laboratory) was a man named Jack Whiteside Parsons who was in love with the idea of Satanism and sex magic. Parsons and a group called, Ordo Templi Orientis (O.T.O.)—the Californian branch of the Thelemites—preformed sexual rituals to invoke the Devil.

He was guided by Aleister Crowley's teaching on rituals and thoughts. It is clear that his mind was controlled by an external source. Apparently he and

Walt Disney and L.Ron Hubbard—to name a few—
liked to gang up like hamsters to have sex and invoke
power from what they believe to be a source of power.
Just a good excuse for them to suck a dick.

Parsons was under constant surveillance from
the FBI because of his involvement with Israel's rocket program and espionage for anyone who could continue his work. It was at the same time Mossad (Is-
raeli Special Forces) was blowing up American movie
theaters.

Parsons died in an explosion in his basement.
Later the FBI found photos of Parsons and his own
mother having sex. His mother Ruth committed suicide immediately by taking a fatal overdose of barbi-
turates. This family was clearly under mind control
and when they awoke from their hell, they couldn't
live with themselves.

One of the other main contributors to NASA
was Wernher Magnus Maximilian Freiherr von
Braun. Braun was a NAZI (National Socialist Zionist)
rocket scientist that was captured and brought to
America in Operation Paperclip to work for NASA.
His application to the NAZI Party was accepted and
his membership number was 5,738,692. In 1940 he
joined the SS, his membership number was 185,068.

Braun developed intermediate range ballistic
missiles, and was the chief architect of the Saturn V
heavy lift launch vehicle that propelled Apollo space
craft. Braun was an advocate of human missions to
Mars. The handlers of this NAZI controls many hu-

mans to this day and will continue this and other missions at all cost—even perform test to see how much hydrogen cyanide a human can survive. In 1975 he received the National Medal of Science.

This is what NASA's foundation is based on. We know nothing of their findings if there are any. NASA was built on lies to beat the Russians into space and claim the power of technological advancement over them. Now NASA poisons our bodies and water supply for the new religion—Science.

Until recently, the only civilians that have ever attempted to enter into space was a teacher and reporter—they both died in the Challenger explosion—no coincidence.

Now we have extremely rich people playing superheroes. Everyone wants to be Tony Stark from the Marvel Comics. Only the people with extreme wealth have the ability to live that fantasy. Elon Musk is blowing up rockets in south Texas as you read.

I wonder what our air and water quality is like on the sites where rocket fuel burns and permeates our soil, and emits soot into our air. I'll bet TCEQ gave Tony Stark a free pass to do whatever he desires in Texas. As long as he brings his check book with him.

The writers and operators of Public Law 105-85 must be imprisoned or destroyed.

How many human bodies have been placed in cemeteries that are time bombs waiting to release a disturbing amount of poison in our water supply? That number is unfathomable.

Embalming fluids are a dangerous contaminate that leaches into our water supply. Most of the time embalming fluid is a mixture of formaldehyde, glutaraldehyde, methanol, and solvents.

Around 5.3 million gallons of embalming fluid are used every year. Formaldehyde is one of the top 10 most hazardous chemicals that leach into our drinking water.

It is carcinogenic (cancerous) in humans and animals and kills electrons leaving the remaining electron desperately trying to survive by stealing electron positions of other cells. It causes the ousted electron to steal any other position for survival. This pattern continues, multiplies, and replicates—we call these <u>free radicals</u>. This isn't the only contaminate that causes free radicals.

Free radical - an uncharged molecule (typically highly reactive and short-lived) having an unpaired valence electron.

You will soon learn about dangerous, compressed, high powered, millimeter (laser beam) frequencies that are indeed ionizing. Ionization is to remove or kill an electron from an atom, and thousands of cellular devices around you day and night cause this to happen.

Radio frequency radiation, especially a microwave frequency, can transfer energy to water molecules. High levels of microwave energy will generate heat in water-rich materials such as most foods. This efficient absorption of microwave energy via water molecules results in rapid heating throughout an object, thus allowing food to be cooked more quickly in a microwave oven than in a conventional oven.

The water molecules in your body are cooked the same, also causing the water molecules to be deconstructed and evaporate causing heart complications and many other dysfunctions related to dehydration. It's always relating to the theft and destruction of our water.

Radio and television broadcasting, cellular telephones, personal communications services (PCS), pagers, cordless telephones, business radio, radio communications for police and fire departments, amateur radio, microwave point-to-point links and satellite communications are just a few of the many telecommunications applications of Radio Frequency energy or RF that destroy the water molecules in your body.

Some bodies are embalmed for viewing then cremated at a later time. Formaldehyde released from the cremation of these bodies are suspended for up to 250 hours after process. The formaldehyde vapors are water soluble and bonds with moisture in the atmosphere—then rained down onto humans, plant, animals, and water supply below. Formaldehyde content in precipitation can range from 110 µg to 1380 µg per liter.

There are stories on the streets that PCP is the same product as formaldehyde (embalming fluid). In fact, there are documented cases where people have broken into funeral homes to steal formaldehyde because it can produce a high resembling that of PCP. However, PCP and embalming fluid are not the same product.

Embalming fluid reportedly produces a hallucinogenic effect and causes the cigarette to burn more slowly, potentially resulting in a prolonged high with severe neurological damage.

Could embalming fluid in the water supply produce a hallucinogenic effect in these doses? Could embalming fluid cause schizophrenia and schizoaffective disorder? Not only visual, but Olfactory (smell), tactile (touch), gustatory (taste) causing a severe alteration of reality in every way.

Our water supply is a delicate living being—it is electric, vibrational, surging with life. In places like Exxonmobil Refinery Downtown Beaumont the vibration is extremely low.

Exxonmobil Refinery Downtown Beaumont has a thing they like to blame everything on called the Oil Plumes. It's a large site under the flares that is covered with asphalt. The asphalt covers the ground to preserve the soil from being disturbed. Apparently in the early 1900's oil was stored in earth bottom tanks (oil tanks without a bottom that sits on the ground). The oil seeped into the top soil causing a large area to become contaminated.

It was one of the first times high volumes of oil was to be stored. They knew nothing of the dangers to the water supply or the hazards of health and life. If they did—they just didn't care.

Today they claim to have French drains that collect the seepage. If any oil gets into the river for any reason Exxonmobil Refinery Downtown Beaumont reverts back to their grandfather clause and blames it on the oil plumes (former site of the earth bottom tanks), and they are relieved of damages. If they get a hang nail—it's the earth bottom tanks fault.

I understand there are broken transfer lines under the Coker unit (located by the river) that they don't want to spend the money to fix. They would rather gush out oil and fuel into the ground and water supply and hope that most of the fuel gets through the line.

Even TCEQ knows about it and does nothing. TCEQ doesn't have the testicular fortitude to stand up to anyone of power. Every once in a while they will go after a small business owner for pollution that happened before they were in control of the business. And they will make an example of someone who can't fight back—even jail time will be served. When it comes to Exxonmobil—they are a team player ready to help. If that's alright with Exxonmobil.

Now if you call TCEQ there will be a long message about Covid and a phone number at the end and the message immediately hangs up. It's to keep you from reporting pollution. TCEQ now has a company called ChemTel Inc. answering the phone—where an undesirable bitch tries to cause you trouble and intimidate you from filing a complaint.

When you finally talk to a TCEQ employee, they, as well, try to intimidate you and try to convince you to not file a complaint. This is a sudden shift if policy for TCEQ. Just a short while ago TCEQ was very helpful—now they are useless as tits on a boar hog, like Greg Abbott.

This Covid bullshit isn't going to work. We will stand not for a Chemical Telecommunications company running our state run entities. ChemTel must be punished for running interference for Governor Greg Abbott—do all you can.

Whoever made the decision to place ChemTel in the position must be punished as well.

These transfer lines and oil plumes must be remediated immediately by Exxonmobil (the Kingdom of Saudi). No more get out of jail free cards. Once this poison is remediated—all other contaminates from Exxonmobil (the Kingdom of Saudi) will be clearly identified and charged.

The Kingdom of Saudi (the owners of Exxonmobil and most all refineries and chemical plants) has an obscene amount of money and could fix all dilapidated equipment in their refineries and chemical plants. That would cut into their profits. The Kingdom of Saudi likes that they can shit on Texas and Americans on the cheap.

People in the oil industry are conditioned to think they can do as they choose because they are more valuable than any other human. When I was a kid, a neighbor's water well was damaged because of an oil rig near by. In retaliation the oilfield workers beat up the family's teenager because they sued the oil company. The oil companies demonize their employees.

Speaking of on the cheap—most harbors don't allow ships to dump raw sewage into the rivers—on the Sabine and Neches you can. Think about that the next time you bask on the river's edge. They are very obedient to money in Southeast Texas.

A whale washed to shore with 88 lbs of plastic in its stomach in Compostela Valley in the Philippines. Ocean life is being destroyed. Eighty million tons of plastic wash into the ocean every year.

When I was a kid there was a lot of ocean life on Crystal Beach. Everywhere you looked there was crabs with big and little claws, shrimp, hermit crab, horseshoe crabs, sand dollars, oyster, seahorses, jelly fish, seahorses, sting ray, and little fish everywhere. Now the beach is void of life.

I have a friend who is an Army Diver who spent many hours under water in the Middle East near oil platforms in the Persian Gulf. He tells me to reach the gulf floor they swim through 10' of trash ranging from plastic to metal and raw sewage.

That's 10' of trash covering a lot of the Persian Gulf floor (Port Shuwaikh, Port Shuaida, Kuwait, and many more). Like most oil platforms world wide, they disregard the ocean and are conditioned to believe they are at the top of the food chain.

He witnessed a common practice which is to dispose of their garbage in the ocean—that's lift the plastic bag from the trash can, and throw it directly into the water. No one will stop them, they are oil men to do as they choose.

This is the reason there are plastic shards in the fish we eat. This is why we have plastic shards in our rain drops. The pain of this petroleum product out weighs the return.

Another major concern of his is in Fort Lauderdale, Florida—a place called Osborne Reef.

In the 1970s over 2 million tires were put into the ocean to create additional fish habitats. Through the years hurricanes moved the tires—as a consequence, the movements of the tires damaged the real existing coral reefs. In addition the leaching polypropylene (PP), polyethylene terephthalate (PET), polystyrene (PS) and polyvinyl chloride (PVC) microplastics.

Florida Department of Environmental Protection finds this threat very serious. It was decided in 2007 by the military to start the clean up efforts, and a pilot program was needed to test drive retrieval productivity, loading and transportation methods, and tire processing and disposal. It's refreshing to know the military is put to good use.

Mexican resorts drive their dump trucks to the edge of the dock and dump their trash directly into the Gulf and ocean—videos clearly show this. It's an act of war against humanity and all living beings.

The Kerala Forest Department in India—the Nilambur North Division—produced a sign that reads "Save our forest from plastic Devil. Plastic free zone; Littering is a punishable offense." In this region plastic is smothering all life and it's growing like cancer.

The President of the Philippines threatened to declare war on Canada if they didn't take back thousands of containers of trash it shipped over some years ago.

No wonder Canada is so clean—they ship their trash to poor countries and have ignorant white trash refine their oil on the Gulf Coast.

Here's the hard truth—<u>petroleum products are too dangerous to be in the hands of humans</u>, and production will come to a halt. We have multitudes of alternatives. Recycling is a way of convincing you to hang in there, we can still use the product. Producing this product will kill us.

How do humans live within the real limits earth provides? We are not the only living being on the Planet and we will not survive without diversity of life.

Even our planet is in the Goldilocks portion of our solar system—where it's just right. Balance of give and take will be inherited. Where my rights end your rights begin—all living beings have the same right as well.

Ecuador is the first country to recognize Rights of Nature in its Constitution. Rather than treating nature as property under the law, Rights of Nature articles acknowledge that nature, in all its life forms, has the right to exist, persist, maintain, and regenerate its vital cycles.

The right of property is an ugly necessity in the human reality we exist in and feels unnatural.

Without the ownership of yourself someone else could own you and, in many situations do. Maybe you don't have physical shackles—there are indeed psychological and emotional shackles. If someone believes they can own property, then they do. The person without the belief of property will be consumed.

Our souls possess our bodies—your body is possessed by you—you are indeed possessed. You're here to control this possession. Without control of this possession your body acts a fool—coullion.

It's like getting rid of all dangerous weapons. If they exist, then taking them away will only leave the person without a weapon vulnerable. It seems counter productive—the trouble with most systems is humans.

Human nature must be accounted for in all systems, and it looks like it has always been and always will be the way humans exist. As long as we inhabit these meat vehicles, these conscious corpses—the reality of person and property will remain.

However, Mother Earth is a living being—with the same rights as you and me. She has the ownership of person and property as well.

Bolivia ensured their portion of earth will be protected. It's called Law of the Rights of Mother Earth. They include: the right to life and to exist; the right to continue vital cycles and processes free from human alteration; the right to pure water and clean air; the right to balance; the right to be free from pollution; and the right to be free from cellular structure modified or genetically altered.

Mother Earth rights will be achieved. I'll even bet that the pendulum swing will—for a very short while—get out of hand and go too far. Eventually balance will be restored.

In 2017, the city of Mexicali, Mexico finalized a deal with Constellation brands. This company based out of Victor, New York produces Corona, Modelo Especial, Pacifico, Negra Modelo, as well as American craft beer producer Funky Budda.

Craft Beer - a beer made in a traditional or non-mechanized way by a small brewery. This sound like a craft beer to you? Or Mexican for that matter?

Constellation Brands, Inc. is a Fortune 500 company and is currently embroiled in a water rights dispute over a proposed brewery in the desert city of Mexicali, Baja California, Mexico.

This water comes from the Colorado River, which not only provides daily use of water to Mexicali and the agricultural valley, but is also piped to the mountain and coastal cities of Tijuana, Tecate, Rosarito and to some areas of Ensenada.

Water is limited in the desert—you would think an educated Fortune 500 company would know that. So exploitation of the desperate and deprived Mexican people must be on the agenda. It's hard to see the pain of the Mexican people all the way from New York. God knows the Mexican government are full of murderers and thieves.

In 2010, a French court in Carcassonne, France, convicted 12 wine traders and producers for selling fake Pinot noir wine to buyers in the US, amongst them Constellation, in a scheme that lasted from January 2006 to March 2008.

Constellation later faced and settled a class action lawsuit because it "should have known" that the wine was fake.

It is obvious that Constellation Brands are not only criminals that make beer that tastes like piss, but also doesn't care if they take more than their share of water for a lesser recreational product.

Constellation Brands enjoy being able to exploit Mexicans and believes their own lives are more valuable than a bunch of border town Mexicans. They would rather these Mexican animals go scavenge somewhere else.

Constellation Brands is much like Nestle—real psychopaths. Nestle produces shitty food and drinks through coercion.

In 2018, a 9th U.S. Circuit Court of Appeals reinstated a lawsuit by a group of former child slaves. Doe v. Nestle, S.A. The child slaves accused the U.S. unit of Nestle SA and Cargill Co. "of perpetuating child slavery at Ivory Coast cocoa farms." The child slaves, originally from Mali, are contending that the said companies aided and abetted human rights violations through their active involvement in purchasing cocoa from Ivory Coast (Côte d'Ivoire).

They originally sued Nestle USA, Archer-Daniels-Midland Co. and Cargill Inc [CARG.UL] in 2005. Archer-Daniels-Midland was dismissed from the lawsuit in 2016 according to court records.

The case is back in the Supreme Court, which rejected the companies' bid to have the lawsuit thrown out. The judges unanimously agreed that the group could proceed with its claim despite the fact that the alleged abuses occurred overseas. Lawsuit docket number 17-55435. Now here's a group of judges that can survive the Great American Purge. Nestle sold to Ferrero in 2018 for $2.8 billion—will the sell prevent prosecution for Nestle criminals?

We must damage companies like Altria (who owns Phillip Morris USA, Kraft, Nabisco), Hershey, Nestle, and Mars for crimes against humanity—pain for profitability. Not only does their product lack quality nutritional properties, they also have to steal human life to keep it afloat.

You ever hear Chevron Phillips say their company is run on human power? There is a lot to be said for the vampiric draining of the human consciousness. Freeze and seize Nestle and alike companies' assets. They deserve to be wiped from existence.

Some believe it's a good idea to place a higher price on water to keep waste and industry in check—this idea is the worst solution. The consumer will be the only party to suffer. The answer is everyone has their share. All living beings deserve and require this life giving source—water.

Yemenis are living in hell. Lake Chad is almost all the way dry—its sand kills ocean life in Florida. Hundreds of thousands of lives have been lost in Darfur since 2003 because of a water war. The Syrian civil war was caused by drought; lakes in America are reduced as well. There's been a 15% increase in cost in Sydney—maybe more since the continental fire. We are on the verge of a real water war even in Texas—where most all Texans have guns.

The nations are angry, and the wrath of the lamb is coming. And them that fear thy name, small and great; and shouldest destroy them which destroy the earth. Revelation 11:18

The meek and the mild are the lamb. When a lamb reaches its threshold it's already too far. When a gentle man becomes violent—his violence exceeds.

This is when the rivers run blood red. We can divert or live this revelation. It's up to you.

Air

XVI

Breath - noun; The power of breathing; Life force. Without the breath you're not alive. Life enters and leaves the body through the breath.

In with your very first breath, out with your very last. The breath of God is life itself. It is the fuel that keeps human bodies from returning to dust.

What being wishes to replace the breath of God with something unnatural?

The air we breath consists of Nitrogen 78%, Oxygen 20%, and traces of the other 2% is Argon, Carbon Dioxide, Neon, Methane, Helium, Krypton, Hydrogen, Xenon, background or ground level ozone, Nitrogen Dioxide, Iodine, Carbon Monoxide, and Ammonia.

However, through monitoring equipment around the world; it is clear that industrial ash, background ozone or ground level ozone (O^3) has greatly overcome the nitrogen oxygen ratio required for survival.

The word Ozone with a capital "O" refers to the Stratospheric Ozone. It naturally occurs in the earth's upper atmosphere—6 to 30 miles above the earth's surface. Stratospheric Ozone shields us from the sun's and cosmic radiation. Stratospheric Ozone has nothing to do with the pollution or smog that is killing us.

The words background ozone or ground level ozone (O^3)—with lower-cased lettering—are one and the same and refers to smog and pollution. These two names that refer to smog and pollution are the supplier of cancer, sudden heart attacks, seasonal flu, viruses (which are exsomes), and a great number of other illnesses. Background ozone or ground level ozone (O^3) along with industrial soot from Mexico and Africa is the cause of what is referred to as Covid.

It is a toxic agent to the respiratory system. You indeed should limit your exposure to elevated ground level ozone or background ozone (O^3) of more than 38 ppm.

I was taught in school that the Ozone in the sky was disappearing and the sun was going to kill us for it—look at the sky and fear it.

It was a clever way to convince children—who will become adult employees—that the trouble was the sun, not refineries, combustion engines, and thousands of acres of burning tires in Kuwait. Keep looking at the hand waving while the other hand socks you in the stomach.

For some of you ignorant readers—when New Zealand says it doesn't have any ozone, it doesn't mean the sun in New Zealand is killing people; it means they lack smog, pollution, and sickness.

New Zealand will try to scare you into not going to their country to preserve their health and I can sympathize.

New Zealand is also a part of the Five Eyes (the world wide surveillance state)—connected directly with the Communist party to control and invade all humans' privacy. It's laughable and they are the enemy of all living beings. New Zealand is indeed Communist and we should invade them.

This is one of the many ways science—the new religion—has misguided and buckled to industry. It is a clear way to downplay ground level ozone or background ozone (O^3) and confuse you.

Ground level ozone and background ozone (O^3)—referred to as smog, is a noxious gas formed in the ground level atmosphere when three oxygen molecules combine.

It is a chemical reaction between oxides of nitrogen and other volatile organic compounds—such as crude oil—when present in sunlight. It comes from emissions from Mexico, Central, South America, Africa, industrial facilities, electrical utilities, motor vehicle exhaust, gasoline vapors, and chemical solvents, oil and gas production, and a variety of other industries.

We have our own pollution flowing into Mexico as well with the winter north wind polluting them.

Tropospheric transportation (air lines) is a major cause of ground level ozone and background ozone (O^3) and a multitude of tumors and cancers.

In the United States, it starts in the spring with the poisonous south wind, and ends in the winter when the north wind pushes the toxic air back to Mexico. However, in the winter months industries such as coal power plants emit a very large and constant heavy sulfur dioxide mass onto all of America from the burning of sulfur or materials that contain sulfur, and indeed covers the entirety of North America causing what is referred to as Covid.

Ground level ozone or background ozone (O^3) adversely impacts growth of vegetation, with a crop loss of $18 billion in the year 2000 alone. <u>It stops the growth of food required for survival.</u>

Crops require carbon dioxide for photosynthesis—humans and animals require clean breathing air to produce carbon dioxide.

Ground level background ozone (O^3) removes and takes the place of these requirements. Crops in 2020 didn't do very well—ask anyone who planted. Ground level ozone or background ozone (O^3) is heavily emitted by fuel combustion from agriculture and biomass burning.

On June 15, 2020, the African dust storm the media referred to as "Godzilla" arrived through the Gulf of Mexico and on to American soil.

NASA's Global Modeling and Assimilation Office released an image of the Dust Aerosol Optical Thickness that showed high amounts of background ozone or ground level ozone to reach Central Texas. NASA - NOAA's Suomi NPP Satellite collected the data and all of the media used the information as the baseline for their reports.

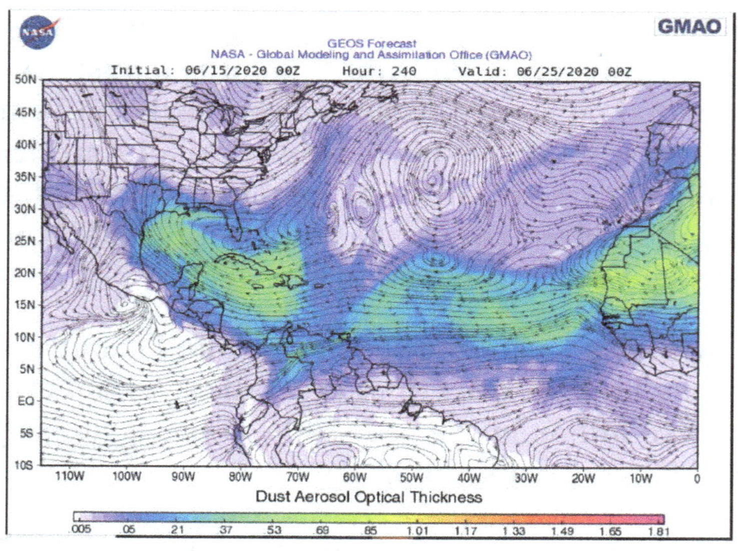

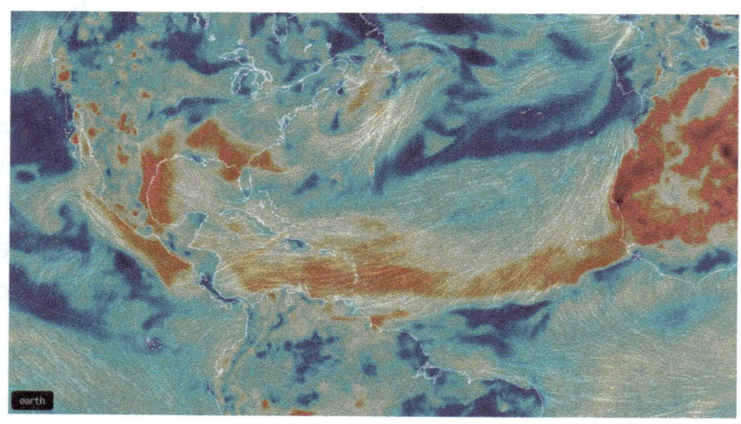

     That same day (June 15, 2020) San Marcos, Texas had 400 cases of the Covid - 19 virus. By June 18, 2020, San Marcos, Texas had 800 cases of the assumed virus. June 16, 2020, there were 155 new cases, June 17, 2020, there were 145 cases, and more than 100 new cases on June 18th.

PBS News Hour reported 44,000 new cases of sickness on June 30, 2020. Sound like a virus to you? Or are we being poisoned and attacked with biological weapons—ground level background ozone (O^3) - by other countries such as Africa and Mexico using the natural wind patterns? It's subtle but effective.

During this same period—thru July 4, 2020—ground level ozone and background ozone (O^3) from Africa caused people in Florida to induce more than 1000 cases of sickness.

By July 8, 2020, there was a 40% increase to more than 62,000 cases in one day. Ground level ozone and background ozone (O^3) was higher than 101 ppm—and still, the media and government officials want us to wear cloth masks?

<u>Ground level ozone and background ozone (O^3) is the removal of breathing air.</u>

Anyone that works in an underground tunnel knows there is a great potential for death due to the lack of breathing air. This is why we follow proper

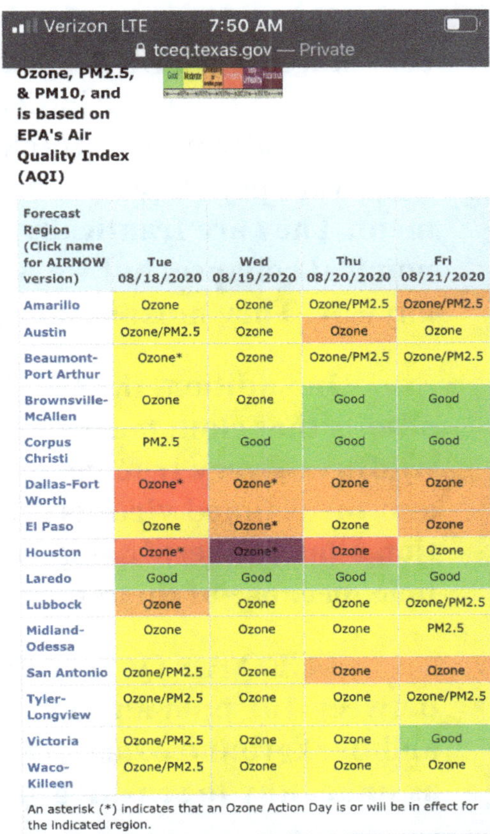

Ozone, PM2.5, & PM10, and is based on EPA's Air Quality Index (AQI)

Forecast Region (Click name for AIRNOW version)	Tue 08/18/2020	Wed 08/19/2020	Thu 08/20/2020	Fri 08/21/2020
Amarillo	Ozone	Ozone	Ozone/PM2.5	Ozone/PM2.5
Austin	Ozone/PM2.5	Ozone	Ozone	Ozone
Beaumont-Port Arthur	Ozone*	Ozone	Ozone/PM2.5	Ozone/PM2.5
Brownsville-McAllen	Ozone	Ozone	Good	Good
Corpus Christi	PM2.5	Good	Good	Good
Dallas-Fort Worth	Ozone*	Ozone*	Ozone	Ozone
El Paso	Ozone	Ozone*	Ozone	Ozone
Houston	Ozone*	Ozone*	Ozone	Ozone
Laredo	Good	Good	Good	Good
Lubbock	Ozone	Ozone	Ozone	Ozone/PM2.5
Midland-Odessa	Ozone	Ozone	Ozone	PM2.5
San Antonio	Ozone/PM2.5	Ozone	Ozone	Ozone
Tyler-Longview	Ozone/PM2.5	Ozone	Ozone	Ozone/PM2.5
Victoria	Ozone/PM2.5	Ozone	Ozone	Good
Waco-Killeen	Ozone/PM2.5	Ozone	Ozone	Ozone

An asterisk (*) indicates that an Ozone Action Day is or will be in effect for the indicated region.

A caret (^) indicates that levels of PM may exceed the applicable short-term procedure from the 40 hour OSHA HAZWOPER Training (Hazardous Waste Operations and Emergency Response Standard). Many lives were lost to understand and accept this philosophy.

If you received this training you would know that a paper mask with the best micron is improper Personal Protective Equipment (PPE). You would also know that a full face respirator with carbon cartridge filtration system would be improper PPE as well.

<u>The only proper PPE required when clean breathing air is removed would be supplied air.</u>

No-one has the money or gall to go to the grocery store with SCUBA gear, nor should they be forced to. It is a clear indication that all living beings are physically under attack by industry—directed by corporate psychopaths and protected by the government. They are frantically trying to stop a world economical collapse, but we won't survive the pollution anyway. They must be stopped by force or killed.

In addition, the CDC accidentally, but openly, admits that cloth masks are no help for smoke from wildfires—very true. So if smoke particles are 10μm and the claimed virus is 0.1μm, then through deduction a mask doesn't help with a bigger or smaller particle, because of the removal of clean breathing air.

TCEQ Air Quality Forecast showed air to be between 101 ppm and 150 ppm ground level ozone and background ozone (O^3)—which TCEQ refers to as unhealthy to sensitive groups.

The Weather Channel showed the air quality index to be closer to 160 ppm ground level ozone and background ozone (O^3) with winds at 10 mph south-southeast from the gulf.

Air Pollution in World : Real Time Air Quality Index Visual Map, aqicn.org, showed hour by hour the ground level ozone and background ozone (O^3) to be upwards of 180 ppm.

And the media and all government officials want you to continue to believe there is just a bad virus or genetic disease killing everyone.

This is operation Mockingbird—where all government leadership and media says the same thing and really has no idea what the cause truly is. It truly shows their incompetence and they are all ill-equipped to lead anyone or they are all being paid or extorted to continue the lie.

Everyone who subscribed to this virus facade is your enemy and the enemy of truth. The only purpose for continuing the virus fasade is to continue polluting. You see, taking away our liberties was a great mistake and will tarnish their name in the history books forever.

On July 20, 2021, I made a post on Instagram under the handle UNCLE_SWEETHEART stating that heavy toxic air was coming from Africa, and you don't have Covid, you're being poisoned. In the post I showed a TCEQ text to my phone showing ozone action in Dallas - Fort Worth, Houston, and El Paso.

That's all that would fit on the screen but there were more.

From July 20, 2021 thru August, I showed TCEQ reports of heavy ozone and PM2.5 (which is industrial soot) and satellite images of pollution, and stated that people would become sick and die from said poison. I'm not Nostradamus, I have eyes to see satellite images of the poison arriving, and people indeed became sick and died. Many of whom I knew well.

I called all rise in Covid cases with the help from many sources like TCEQ reports, satellite images of toxic air pollution. TCEQ will at least give you enough truth to prevent them from being held liable. This is a clear indication we are being democided with pollution to prevent a world economic collapse.

Public Law 105-85 allows chemical and biological weapons testing on civilians without explicit consent. The writers and operators of this Public Law must be destroyed.

The fact that we let industry take us to this level is a clear indication of denial, psychological conditioning, and symptoms of Stockholm Syndrome.

Stockholm Syndrome - feelings of trust or affection felt in many cases of kidnapping or hostage- taking by victim towards captor.

Why is the dust cloud from Africa so monstrous?

Because of oil exploitation and Canadian mining companies funded by Canadian International Development Agency, whom operate without any regulation at all. The Chinese are major mining polluters as well. Some African dust has always fertilized and fed the Atlantic Ocean—it's not as much of a problem as the extreme amount of ground level ozone and background ozone (O^3), industrial ash, or the amount of dust we now acquire.

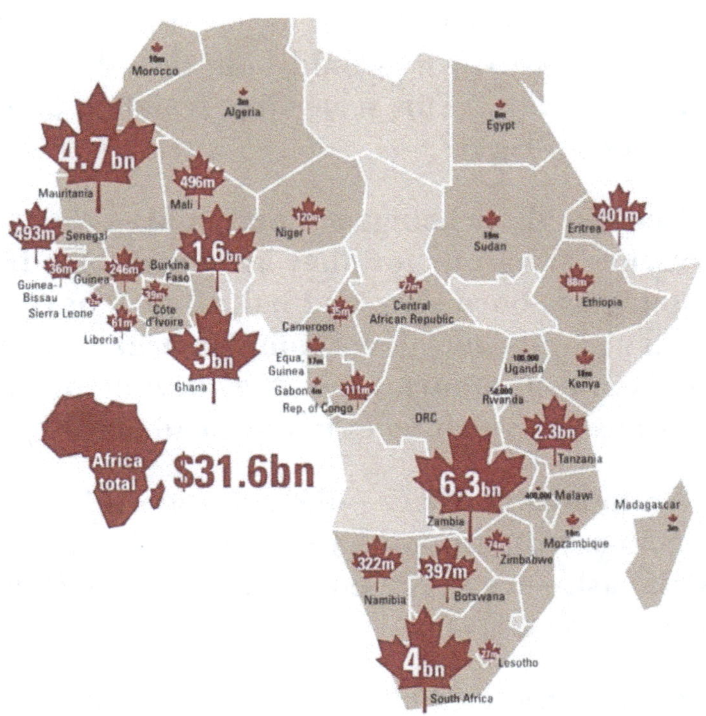

Not only have they removed all of their coolant system—oil—which allows the core of the earth to cook and evaporate what water, moisture, and destroy what vegetation they had; they also exhaust toxic fumes when refining. The only environmental controls they have in Africa are machine guns, which allows them to emit as freely as the chose without complaint—much like Odessa, Texas without the gun.

<u>The more we remove of our coolant system (oil), the more desolate and bereft our planet will become. Continuing to remove oil will indeed cause the earth to be destroyed by fire from the core.</u>

More than 136 Dolphins died on a West African beach in September 2019. Shortly after 26 pilot whales died off the coast of Georgia—the second time for the season. Now why would I put this information into the air section? This seems like it should be in the water section—right?

Because these beings are mammals—they have the presence of hair or fur, they are warm-blooded, their young are born alive and naked, they have mammary glands, they have complex brains (some of us do), and most importantly—they breath air like you and me—through holes in our head.

From mostly Africa—the pollution is killing all living beings with ground level background ozone and industrial ash—it's the removal of clean breathing air over the entire Atlantic.

Easter Sunday 2020 North America was engulfed by deadly ground level ozone and background ozone (O^3). Real time monitoring equipment, aqicn.org, showed readings of over 900 ppm in Aguascalientes, Mexico.

It came from the corner of Avenida Josē Maria Chāvez and Avenida Sigio XXI. Also from Monclova, Mexico at the corner of Calle Mērida and Calle Bogotā.

It was concentrated enough to make it very unhealthy for sensitive groups up past Tennessee, the Carolinas, and down to Florida—winds from southwest 14 mph.

More than 2,000 miles of poisonous air—more so closer to the source. It was above 200 ppm ground level ozone and background ozone (O^3) and TCEQ said it was <u>dangerous to life and health in Port Isabel, Texas.</u>

Mexico has many sites like this and it poisons more than 2/3 of North America.

Just think of yourself in Atapan or Durango, Mexico—where the pollution is at 900 ppm ground level ozone and background ozone (O^3) - with a newborn, fragile.

Just think of where you would go to escape this poison to try and save your own newborn baby—without means or knowledge of how to accomplish this feat. You would have to travel over 1000 miles of hostile territory with an infant and no money. This is an outrage and must be rectified.

Mexico is polluting the air of most all of North America and doesn't care about the health and well being of anyone—including it's own population. It is obvious that the controlling forces of Mexico, Central,

South America, and Africa are psychopaths and are in need of eradication.

Mexico, Central, South America, and Africa have been psychologically overcome by the CIA, Russia, and Communist China—which has military bases with armed Russian soldiers in South and Central America.

Why are road signs in Mexico displaying the Chinese language? The Communists have a lot of money and buying power due to slave labor, fear, and repression. They are also a major land holder in the United States and Mexico.

Let's not forget the Central Intelligence Agency —developed by Wall Street and the Federal Reserve (a privately owned bank)—which is an expert in destabilizing countries and economies and has unlimited tax funds through the CIA Act of 1949 or Public Law 110.

This Act allowed the CIA to use confidential and administration procedures without over sight and are exempt from the usual limitations on the use of federal funds.

This Act also exempted the CIA from having to disclose its "organization, functions, officials, titles, salaries, or numbers of personnel employed <u>without discretion or over sight from anyone.</u>"

Article 1 Section 9 of the United States Constitution, states that, "No Money shall be drawn from the Treasury, but in Consequence of Appropriations made by Law; and a regular Statement and Account of Receipts and Expenditures of all public Money shall be published from time to time."

We must immediately cut off all funds, destroy the CIA Act of 1949 (Public Law 110) and if the CIA puts up any fight we must physically and forcefully stop them.

It was reported on December 22 - January 10, 2018 that Beaumont, Texas was the number one place in America with the most cases of flu. This is after hurricane Harvey—a time of extreme pollution.

All of the plants that had any damage used it as a great excuse and had more needed opportunity to emit toxic fumes. I knew personally more than 50 people who died of respiratory infection and double chemical pneumonia - some healthy athletic humans. I stopped counting at 50.

The same time—after hurricane Harvey—driving between tank farms on West Port Arthur Rd (FM 4513) that belong to Motivia and Valero, I felt a sense of confusion and despair. My lungs became cold and my brain became numb. Numb as in blank—I had trouble functioning or having the cognitive ability to get to my destination.

Around the corner—bordering—are the Mark Styles, Gist, and Le Blanc prison units.

It smelled like the additive they put in the cooling tower at Exxonmobil Beaumont Refinery and

Chemical Plant—supposedly an anti-legionnaires disease concoction.

I wonder what the ramifications will be of this mist emitting into our already humid air with ground level ozone and background ozone (O_3).

Are they involuntarily inoculating us from legionnaires disease without consent? The container this chemical comes in says it's dangerous. Will we receive an outbreak of legionnaires disease like in Flint, Michigan in 2014-2015?

This poison is killing and damaging human life in and around these prison units. Not only the prisoners, but the guards, and staff. It is causing the inmates and guards to act more erratic, irrational, and volatile.

These fumes provide a deeper level of pain which is counterproductive to rehabilitation. The staff's work performance and humanity is replaced with mechanical drone like actions. This and all parties are suffering from these inhumane conditions.

No one can be rehabilitated or rehabilitate when living in constant subconscious fear and anxiety. Being imprisoned means to restrain someones liberties—poisoning them is murder.

Any time the State of Texas, or land owners of the marshlands have a controlled burn, all refineries and chemical plants burn off all the waste they have stored. They all wait for cover like smoke, fog, or storms. Anything that will provide cover for the pollution being seen—3 am is good cover as well—most all are asleep. This has caused many illnesses and deaths.

CenterPoint Energy has a monopoly over our natural gas supply in East Texas, South Texas, Southwest Louisiana, and Arkansas . There are gas leaks all over Beaumont, Texas that they don't give a fuck about. If you try to report it, they want to know if you have service. If not, go fuck yourself. They preach safety, but it's obvious to everyone in town that public safety is the last thing on their mind.

It can distort your perception, and bring about a deeper fear and anger. Ground level or background ozone (O^3) can separate you from the source—causing great mental anguish. Ground level or background ozone (O^3) makes thought and emotion more intense.

It is easy to get a crowd riled up when they are in constant fear and anxiety of dying from the lack of clean breathing air—especially if you electrify the heavy metallic soot constantly in the air.

Most of the people at the January 6, 2021 D.C. protest were from Pennsylvania and are all inundated with a massive amount of sulfur dioxide coming from York, Pennsylvania—steel country.

Sulfur dioxide is released into the air from the burning of sulfur or burning of materials that contain sulfur. Check the site classic.nullschool.net, click on the Earth tab, click Chem, and then SO2sm overlay. It will become very clear.

The release is unnecessary and is easily fixed with pollution controls like citric acid in wet scrubbers. The companies that release the sulfur dioxide can't profit with the extra cost because of the fasade of the fixed Capitalist market; however, we also can't survive the pollution.

Sulfur dioxide, metallic soot, ground level or background ozone (O^3), and the electrification of the metallic soot from the 5G technology can infuriate anyone—another symptom can be anxiety.

It helped that the cult of QAnon misguided them. I say they are no more guilty of aggressing the capital than the companies releasing the sulfur dioxide.

On December 10, 2021 a massive and dangerous tornado developed from the sulfur dioxide surface mass—killing many in Kentucky for two hundred and twenty seven miles. The comparison for Texans, that's the destruction of everything from Beaumont past Austin, Texas.

Sulfur dioxide causes weather to become extreme and creates dangerous winter storms. Every winter, there is a shrimp whisker of sulfur dioxide (created in the Tennessee Valley and the Appalachia) that extends from New England to damn near Iceland. This sulfur dioxide shrimp whisker joins the same two developed polar vortex to create one giant, severe, and deadly winter storm—every year. You can watch this happen in real time on classic.nullschool.net—every year.

Sulfur dioxide is created by the burning of sulfur and/or the burning of products that contain sulfur (coal) to create an abundance of electricity for heat in the winter. It's a vicious cycle of needing electricity to create heat for the winter—and creating electricity with the by-product of sulfur dioxide that creates powerful winter storms—insanity.

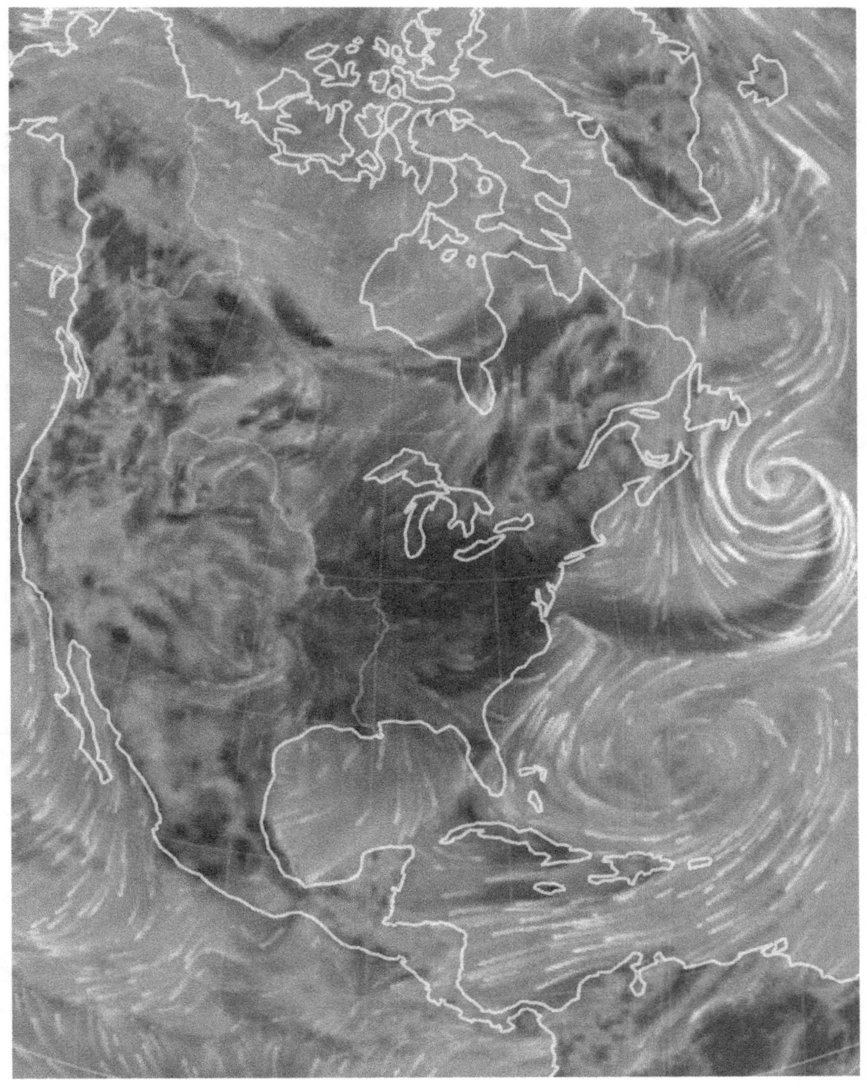

Everyone is being constantly attacked by ground level or background ozone (O^3), metallic industrial soot, sulfur dioxide, and carbon monoxide and all components reduce the ability to rest.

The ground level or background ozone (O^3), metallic soot, sulfur dioxide, and many other poisons we breath causes your injuries to become inflamed. Inflammation is big business and one of the main symptoms that all living beings suffer.

When your injuries become inflamed check the air quality site I listed and see if it coincides—you will be astonished. Poisonous air somehow makes you want to exercise more on toxic days.

Is it the human condition that makes the body desire poison, or is there psychological conditioning triggered to cause you to destroy yourself?

Everyone has to labor more to stay alive—more so than designed. We can progress much faster when we have clean air and water. Our cognitive ability will grow; our bodies will become bigger and stronger.

I guarantee people who mass shoot were driven to it with the help of ground level or background ozone, wireless technologies along with social media—to die and take people with them. They can't do it anymore and figure they can go out eliminating their perceived enemy—giving themselves up as a sacrifice like Jesus.

The ground level background ozone (O^3) in Southeast Texas, the Gulf coast, and across the State of Texas makes people hyper-aggressive with high anxiety.

Think of all the anti-depression drugs being prescribed to most everyone you know. The opioid crisis developed with the help of anxiety from the lack of clean breathing air—the most addictions are the most polluted. <u>When the ground level or background ozone (O^3) disappears so will your pain, anxiety, and depression.</u> Check air quality sites to see if your depression or anxiety coincides with poor air quality, then you'll know.

Even though we had a world-wide shutdown due to the 2020 pandemic with most humans not driving, and oil markets stopped any extraction of crude —we still had 113 Ozone Action days in Texas alone. That's TCEQ's claim of ground level or background ozone (O^3) higher than 151 ppm and considered unhealthy.

July 12, 2019

Dallas - Fort Worth	-	**35 Days**
Houston	-	**31 Days**
El Paso	-	**30 Days**
San Antonio	-	**7 Days**
Beaumont.	-	**5 Days**
Austin	-	**3 Days**
Tyler - Longview	-	**2 Days**

July 12, 2020

This was 365 day count. We continued to receive poisonous air after July 12, 2020 from Mexico and Africa in the 2020 year.

Dallas - Fort Worth	-	**16 Days**
Houston	-	**15 Days**
El Paso	-	**20 Days**
San Antonio	-	**3 Days**
Beaumont.	-	**4 Days**
Austin	-	**2 Days**

And the winner for the most days of killer pollution in a 365 day count plus the additional measurements days is Dallas - Forth Worth, with El Paso trailing just one day to come in second.

Don't let these numbers fool you. The State of Texas keeps a strong and constant level of ground level or background ozone (O^3) and metallic industrial ash.

In hind sight it would have been easier to count the days of clean air Texas received. I know that you will be helping me count those clean air days for the following year.

Shutting us down only prolonged and will intensify next year's wave of toxicity. Every time government tries to save us, they only make it worse. We needed to feel the pain of dying from the inability to breath in order to correct the root cause.

The State of Texas doesn't have enough money to pay us to not drive every spring and summer. They know what the problem is—leadership lacks the testicular fortitude to correct it.

It's an embarrassment and an indication they lack a brain and backbone. <u>We need people who can make the hard decisions that truly benefit all life—animate and inanimate.</u>

Barges flow down the Intracoastal Waterway emitting toxic emissions like benzene fumes. They vent barges to keep from expanding and exploding—

but they do nothing to capture these carcinogenic fugitive emissions.

Some people think they can escape the pollution by moving to Galveston, Texas or Crystal Beach when they are indeed receiving higher doses of ground level or background ozone (O^3).

Texas City is polluting the hell out of that region from across the bay. Carbon monoxide (CO) is heavier than air, lighter than water. It sits on top of the water surface and it only takes a few breaths to kill you while boating.

It can easily travel on open water—gulf, bay, etc. No-one has the ability to keep track of off-shore oil rigs because of the amount of oil rigs, the miles between them, and most are owned by foreign countries—China, Russia. Not only are they polluting the Gulf with millions of gallons of crude oil—they also throw their trash off the edge of the rig like throwing it in the dumpster.

On March 21, 2019, the National Guard was called out to help residents escape or shelter in place from, at first, dangerous levels of benzene being released by one of the Mitsui & Co. facilities called Intercontinental Terminal Company—which later caught fire—blanketing Houston with toxic smoke.

The Texas Commission on Environmental Quality said it was no big deal—never mind the nose bleed, nausea, or painful headache—but be sure to hide in your tub from the potential blast.

In many cases, people of Southeast Texas subconsciously escape to Austin, Texas because it was the first stop to clean air and civilization.

Houston, Texas or Southwest Louisiana is no safe haven from toxicity. You must travel crosswind to relieve the pain of ground level or background ozone (O^3)—cross wind from the Gulf of Mexico, which is an expressway for toxic air from Mexico and Africa. It runs faster with vastness.

Apparently there's nowhere to run because the air quality is bad all over Texas and America. Nothing but ground level or background ozone (O^3) all summer and most of the year.

High amounts of carbon monoxide and ground level or background ozone (O^3) are on Barton Springs Road and South First, Congress and River Side, and many more places like this in Austin, Texas. If you can't quite understand what ground level or background ozone (O^3) smells like or feels like—just go to the bottom of the hill to the Whataburger.

The ground level or background ozone (O^3) makes homeless people more erratic on "Ozone Action Days" issued by the Texas Commission on Environmental Quality.

Homeless jumpers from cranes, ramped stabbings of tourists, beatings of tourists and brutality from the homeless is city wide. This is why the Mexican people are being slaughtered by the government and the cartel—they have all gone criminally insane due to a massive amount of pollution causing them to become extremely barbarous.

A massive amount of homeless live under overpasses in Austin, Texas for shelter. However they are also addicted to the ground level or background ozone (O^3).

Harmful drugs actually poison you to induce dopamine to release into your brain. That is how you get high, dopamine is the actual drug, the substance you ingest is just the catalyst for the release. The dopamine is what addicts are looking for. This is whats happening to our homeless brothers and sisters living in constant ground level or background ozone (O^3).

Buildings are concentrating the carbon monoxide at the bottom of these hills with I-35 in close proximity to the these locations. The over populating of Central Texas is a clear sign of greed and ignorance by local leadership.

We must dush all local authority and send them back to where they came from immediately. I want them ran all the way out of Texas—by force if necessary.

People think they have severe allergies and there are hundreds of variations of allergy medication—it's a multi-billion dollar industry.

The next time you feel sick or think your allergies are bothering you, go check TCEQ Air Quality Forecast (Texas Commission on Environmental Quality), aqicn.org, and the Earth wind map at earth-.nullschool.net for wind direction and concentration.

You will then know where your pain derives from—believe what you see or suffer. If you send a nasty e-mail to the Weather Channel they will supply you with the air quality index on your phone application. It will be 20 points off actual index—if they say air quality index is 50, its really 70 ppm. Weather Channel is poor quality. All air quality sites I list can become contradictory at times—we must have sound information for survival.

TCEQ Air Quality Forecast (Texas Commission on Environmental Quality), and aqicn.org have been very wrong many times as well.

Aqicn.org (non-profit air monitoring) has now taken severely polluted sites in Mexico off of real-time view like Durango and Atapan. Are they withholding information by compliance or force? Good luck contacting them.

I speak a great deal about Texas Commission on Environmental Quality (TCEQ), and how they enable you with information. They are far from the desired mark.

After 17 years, EPA settles racial discrimination case against TCEQ—the Texas Observer reported. TCEQ agreed to install air monitors in a mostly black comunity near Exxonmobil Refinery Downtown Beaumont, Texas.

In the 1990's Reverend Malveau of Charlton - Pollard neighborhood learned of the overwhelming rotten-egg smells (Hydrogen Sulfide) coming from this refinery. People suffered nausea, dizziness, and burning of the eyes. TCEQ did nothing because the State of Texas doesn't care about a bunch of poor black people, they care about revenue.

<u>Where my rights end, your rights begin. That mean no one's—including companies—rights overlap, impede, or aggresses on anyone else's person or property.</u>

<u>Everyone has the absolute right and ownership of them selves. No one has the right to aggress on your person or property—even toxic fumes. These rights are given to you by the source of all—the All Father. No one can administer or take away these rights—you are born with them.</u>

Hydrogen Sulfide is a colorless gas that is corrosive, flammable, and very poisonous. It takes the place of clean breathing air because it is heavier. The threshold of exposure is 0.01 - 1.5 ppm (parts per million)—you could fall down dead in your tracks at 10 ppm, these people get waves of 20ppm - 100 ppm every day of their life.

I know personally because I professionally monitored Exxonmobil Downtown Beaumont for this kind of poison, and reported to all contacts at Exxonmobil. I know where the sources are—so does Exxonmobil. Exxonmobil employees are suffering as well.

In 2018 Lamar University had a social justice symposium and invited a young lady from Yale University to speak on her findings of the Beaumont air quality study she was conducting.

The symposium went on for hours and there was 16 people speaking—the last to speak was this young lady. Lamar University gave her 10 minutes to disclose her findings—not much time compared to all other speakers.

She expressed that there is a severe amount of toxic pollution coming from Exxonmobil refinery and Exxonmobil employees sabotaged and damaged her equipment. I know this is true because I still worked at Exxonmobil at the time.

People bragged and felt superior of their deeds at the refinery. I know because I was there when they said it. Not just one person—everyone laughed at the refinery. Exxonmobil employees are psychologically trained for superiority over everyone including the laws of Texas. Really they're just a bunch of low-paid rednecks.

Some old buzzard from the Development Board feebly defended the air quality in vain. I had communication later with the young lady and I hope she can shine some light on our suffrage.

Lamar University's inadequate opportunity for this young lady is obvious because Exxonmobil Refinery owns Lamar University—and an educated man knows not to bite the hand that feeds them.

On Thursday January 21, 2016, Lamar University had a campus closure because of close proximity and poisonous fumes emitting from Exxonmobil.

The document stated: "In abundance of caution due to activated flares at nearby chemical plants, Lamar University activities are suspended and the campus will be closed for the remainder of the day. Once it has been deemed safe to return another notice will be provided. Cardinal Village Dorms shelter in place until further notice."

This is one of the many times pollution has overcome this area since the start of Lamar University. Nylon pantyhose stockings used to melt off women in the older days—ask around.

Everyone is psychologically conditioned to comply.

Over 100,000 people have recently died in Serbia, Georgia, and Bulgaria from a massive amount of industrial metallic soot and the lack of clean breathing air. Bulgarians are dying by the thousands from pollution from industry—as is Serbia, Georgia, and other places that live under authoritarianism.

Hypoxemia - a decrease in the partial pressure of oxygen in the blood. Hypoxemia reduces the level of tissue oxygenation. Everyone in the world is becoming oxygen deficient in a biotic environment. If you Cross-fit or run, your chances of becoming sick or dying has exponentially grown. When you hear of healthy athletes dying from Covid, Hypoxemia is most likely the case.

Cyanide poisoning - cells are unable to utilize oxygen despite having normal blood and tissue oxygen levels. Cyanide gas was used in the Holocaust.

Do you want to die from the lack of breathing air?

Brazil's Amazon Rainforest produces 20% oxygen for the world and stores carbon dioxide. The burning of the rainforest has put out 228 megatons of CO2 in 2019, and more than 36,000 fires have been started in 2019—that's according to Brazil's National Institute for Space Research.

The INPE (Instituto Nacional de Pesquisas Espaciais) estimates 2,300 square miles of deforestation in 2019 alone.The last drought the Amazon had was in 2016—these fires were started by people—all was enabled by the new policy shift from Brazilian President Jair Bolsonaro.

More than 40,000 plants and animal species dwell there and the deforestation laws are not being enforced. President Bolsonaro fired Ricardo Magnus Osorio Galvao—the head of the INPE—for reporting satellite imaging of this destruction.

The people of this region have no opportunity to make a living for survival other than destroying the very mechanism that nourishes and sustains our life.

Why should the Amazon be the only saving grace for the world? The time is now for the world to have many "Amazons" producing clean breathing air strategically around the world for survival.

The humans in the Amazon have a right to survival as much as the plants and animals. What can we do to have a healthy balance of life and liberty for all living beings in the Amazon? Why are humans only compensated for the destruction of Mother Earth? Try holding your breath while you count that fiat money.

If Brazil could provide safety for eco-tourism, more people from around the world will visit and bring money with them. The trade deal—beef—with the EU is the driving force for the cutting and burning of the Amazon. The European Union is eating too much beef—the need for more grazing land and irritable bowels proves it.

If the Amazon is the lungs of the world, then the residents need to be compensated by the world to leave it as nature intended. They buy groceries and have homes as well. Every human has to work for survival.

<u>I ask again, what being pays us fiat money to exterminate ourselves?</u> The only way to be compensated with fiat money is to destroy Mother Earth, and to allow the core of the earth to cook and evaporate what water, moisture, and destroy what vegetation we have.

Our ability to defend ourselves is damaged and has weakened our position on the southern border. When they smoke us away from our boundary there will be no one to defend and hold ground on the border—other countries and forces will attack. All aggressors are weakening our bodies with pollution so we can't defend ourselves. Texas and the United States is indeed being passively attacked with biological weapons—ground level or background ozone (O^3) and fine metallic soot as a preliminary attack from the continent of Africa. We must destroy the source of this death cloud and its developers.

Everyone must admit there is a problem and an addiction to the very things that poison us before we can correct it. We are indeed at rock bottom.

Show and Tell

XVII

I come from renegade Indians that escaped from the reservation in Oklahoma, stole Calvary horses, and hid out in the Big Thicket of Louisiana and Texas.

The creator has always shown my line of people the truth and we are charged with knowing and acting if we can. We are to see, believe what we see, and correct the wrong within Universal Law.

Having being charged with tasks, we also have demons that disrupt our tasks by attempting to steal or interrupt ours and other's consciousness. These demons bear false witness as a maneuver to halt cooperation of anyone who will listen and believe. Most will believe the first story they hear—which is disturbingly common.

My mother had three brothers and seven sisters and most were sent to an Assembly of God residential children's home. It was located in Fairfield, Texas, north of Houston, and was called the Daniel Children Home circa 1957. The Purcell's grandmother's connection with the Assembly of God is how my family came to know this place. It was a strange time when native children were sent away to religious homes from direction and pressure from religious groups. The saying was "kill the savage—spare the child."

The Daniels lived in downtown Fairfield in a big Victorian home and appeared to be well off. Everything seemed benevolent—wealthy, charitable people taking time to care for neglected children. The fact is, when beings condensed their pain; vultures perch. It's like chumming the waters with pain.

A few of the Purcell children were being abused physically, mentally, and emotionally by the staff of the Daniel Children's Home—not the Daniels. The separation of the children at the home, especially their 2 year old brother, was detrimental as well.

Everyone's name started with brother or sister—such as brother Gary. They dubbed him Blubber Gary as an act of defiance only a child could provide.

The Daniels would send the children out to other families during holidays like Christmas where the children would receive gifts and love from strangers. When the children were sent back to the orphanage the gifts would go to storage awaiting resale as profit for the Daniels.

A lady in Bridge City named Mrs. Stern—who was benevolent and indeed cared and provided for my mother's family—sent gifts and silver dollars to all the Purcell children. None of the silver or gifts were ever received by the children.

The woman pastor of the on-site church beat two of the Purcell girls until one of them—my aunt Debra—bled. She would't cry—she just took it in defiance. My people react not to pain and violence. Death will come before submission.

Some people can't be whipped into obedience, they have to be convinced things are good for them and shown. It did cause her and the others great mental anguish they suffer to this day. The Daniels are responsible and are guilty of aggressing on the Purcell's person and property.

One day my Aunt Joan (13 yrs old)—who had a friend that lived close to the courthouse—walked across the street and asked to speak with a judge.

Not long after she exposed their treachery (1961) the Daniels donated the children's home to the Assembly of God.

The Daniels gathered the Purcell children together in a room and told them, "You are evil and going to hell. You're the cause of the trouble." Just a little more pain before they go. They are Emotional Vampires and needed one more feeding before departure.

The children's home is now licensed by the Texas Department of Protective and Regulatory Services. Today, The Pleasant Hills Children home is under the direction of a Board of Directors. I'm sure they provide good care while under the eyes of Texas. Not all people and children's homes are malevolent.

Years later, my mother and a few sisters attended a wedding in Fairfield, Texas. They stoped off at the children's home and talked with the staff. The staff was surprised the Purcells turned out well.

My mother explained to them there was a time when this place was detrimental to every child that was placed here.

My Aunt Joan had the courage to tell the judge they were going too far with disciplinary actions. She told them no matter what the consequences were.

What you may not know is the demons attack after the whistle is blown. The rich religious connection will pour poison into people's ears in retaliation for exposure for years to come. Once you place a story in someone's consciousness, it's fact until it can be clearly seen. Being clearly seen almost never happens.

When I was three or so my mother brought me to a doctor in Orange, Texas by the name of Jones for a pediatric visit. She had no more left the room for minutes to handle paper work and payment when this demon put his lips on my three year old pecker.

At first I didn't know what he was doing. This was my first encounter with a sexual act ever—I had only been born just three years earlier.

In my consciousness I knew he was violating Universal Law. In an attempt to halt this action I put my sucker—that his secretary gave to me—in his hair.

My mother came back into the room wondering how that happened. While she tried to help him remove the sucker from his hair, I pointed at him and then my pecker over and over.

Jones was terrified and ran out of the room—didn't even get his coat. He got in his car and sped off. His office staff didn't know what the hell had happened. Being new and starting from scratch in this reality, I considered it a defeat on his part—so I thought.

I'm sure after his office staff finally got an explanation, he painted a picture of a child from the 1976 movie "The Omen." This was the start of the attacks from many more demons that identify themselves as combatants of evil such as clergy and members or claimed members of the Masonic lodges.

Who could question a doctors diagnosis? To society he's damn near god—his white coat proves it.

I know this to be true because in this reality your enemy has to tell you what they're doing to you. It's up to you to listen and believe them.

In later years, Masons explained to me their proud moments of destroying evil—not realizing it was me that had received this vile disgust.

Even if it's to call up a child and tell them they were going to kill me. Ten years goes by fast—in their mind they were proud to defeat the Omen child that is perceived to still be a child. Except I was now someone they knew and could trust whole heartedly—I said nothing.

They would have a child call as if it were my class mate, then get on the phone and tell me all the ways they were going to aggress on my person and property—kill me.

As I got older I asked them to meet up to pre-form these violent actions they speak of. No one would confront face to face. They are hidden cowards—hiding, talking, and bearing false witness. Much like Satan—to give half truth in a web of anguish. They don't know God nor does God know them.

At the time it was unclear to me how they could produce small details like the place or action I performed—and then change the story to fit the disgusting narrative. Someone was pouring poison into everyone's ear—enough poison to enrage them and make them perform actions they wouldn't normally perform.

For example, when I was a child, my brother threw gas on a lit fire. I had to slap his ear and side of his head to put out flames on his body, then hit the end of the gas can that was on fire with a stick, burning from the fumes. The story became I lit my brother on fire and hit him with a stick. No one was around to see to confirm.

What I can't quite understand is if a Mason has the technique to control to their own individual benefit, why do their families fall apart? Because repression leads to excessiveness or another external being attacks them and everyone around for calling the powers from the deep dark.

I say Masons are psychologically conditioned to do something they're ashamed of as initiation. Later they would do anything to make amends by doing good in the world. This is theft of consciousness, entrapment, and deception. These tactics are a snare to do the bidding of the controller.

Around five years later from this doctor's visit, an aunt who was on drugs took a few cousins and myself to see some cows. It was a ranch off of highway 62, the next ranch north of my grandfather's place.

All boys, we ran around like heathens climbing on shit, throwing rocks. Then we settled down and went inside the unfinished cabin. We talked and looked around, maybe all got a coke.

One of my cousins was playing with a small hydraulic jack and unscrewed the sustaining screw and a little hydraulic fluid got on the plywood floor.

The rancher named Hamm acted as if it were a big deal, so did the rancher's son who was in high school. "Uh - oh. You better come in here," said the rancher. Meaning you broke my hydraulic jack, now come into this private room to receive your punishment.

Looking back I can remember my cousin's eyes were very surprised when he came out of the room. When we went back to my grandmother's house he had trouble having a bowel movement—he was in pain. We laughed because we were mere children and had no idea what had happened.

Not long after, I ended up back at that ranch with a different aunt. This time there was another young man who played the guitar along with the rancher's son. He seemed to be held captive. The rancher claimed it was his other son—he flat out told me he was no kin to these people. While in the back bunk room, the rancher's son quietly threatened him for exposing this information. He just kept telling the rancher's son he wanted to go home. The rancher's son was acting like a handler for the rancher.

Strangely enough, some of my female cousins, who were in high school with the rancher's son, appeared in the window of the bunkhouse. They lived two properties south. Sneaking and hiding, they were trying to help the young man escape.

The rancher's son ran outside and caught them. I heard them scream—at the time I thought they were friends and were playing jokes and trying to scare each other—I had no idea. Nor did I know what happened to this other young man after I left. I could see from the bunk room that the rancher was violent towards my aunt and refused me in a way I didn't understand.

A short while later my family and I moved into a shell of a house that we were going to build on as we lived in it. Some of our family came to help us move in. The girls that were at the ranch came as well—and they had the rancher's son with them. They looked very concerned and something tells me he wasn't invited.

There was a lot going on and I can't remember any details while he was there. What I do remember was when it was time to leave.

I was in the back of my uncle's white truck playing and the rancher's son walked up to the edge and wanted to talk with me.

"You better not tell—I know where you live." The look on my female cousin's faces was fear. I laughed and asked him "tell what ?" At the time I didn't know what the hell he was talking about. He tried to stare me down—I just kept laughing because I thought he was just messing with me. He was standing on the ground and I was sitting in the back of the truck and I put my hand on his shoulder because he was at the right height.

He bit my hand and picked it up with his teeth —holding only my skin.

I grew up tough so it didn't bother me—I just kept laughing. I just thought he was playing rough.

Some family members came up joking and convinced him to go into a building full of wasps—if he was tough enough—in his mind he was. He went in with a broom and a rolled up news paper on fire. When he went in the building they ran over there and threw a bug bomb into the building and held the door closed so he couldn't get out. He cried and begged to get out.

Just a short while in there—long enough to induce fear—they opened the door and grabbed him up.

They had him by the throat and threatened him and anyone else who fucked with any of our family. He cried like a little bitch—and I laughed because he wasn't very tough.

Not long after this event, my grandfather's house—where my female cousins lived—burned down.

When I was about sixteen I recognized the rancher's son at the Wal-mart in Bridge City. I didn't know to have bad feelings for him—but he was scared.

He was with his new family or at least a lady and a child. I said his name, smiled and said, "Don't you remember me?"

He was afraid and wanted to change the subject. I let him leave because he didn't want to talk anyway.

About five minutes later this son of a bitch caught me in the front corner of the store and pulled a knife on me. His eyes were glassy like he had been taken over by a demon. I know because I have seen these eyes on many more devils. It was a big folding Buck Knife—brass ends with a wood handle.

I smiled and asked him "How would you like for me to take that knife away from you and stick it up your ass?" He was in fear but came at me anyway. I took the knife away from him and he ran. He was knocking over shit to get away. I laughed my ass off and I believe I still have the knife.

I know what you're thinking. Why would someone laugh at someone who wanted to kill them? Two reasons—first, people from out of the blue attempted these actions all the time. So the fear of it was desensitized. Second, the lord reduces their performance and allows me time to act as I choose. They need fear as fuel to fully commit—there was none to be had. I never really wanted to hurt anybody so they were all let go—even when their death seemed imminent.

Also, in my mind; I was looking for the head of the snake. I wouldn't waste my time on a tick turd like him. I would remain quiet until the lord gave me the opportunity to extinguish them all.

Since then, I have realized that the first aunt sold my cousin to the rancher for money and drugs. The other aunt tried to sell me, but they knew I would tell and put up a fight—the network already notified them I would. The rancher almost killed the second aunt and scared her to death. Later she worked at the only local gas station and told everyone she came in contact with that I was a psycho-killer child molester as deflection from her actions—psychological projection. She tried to convince people to kill me to get rid of the evidence that I now express to you.

I ask these people if the aunt ever stole, lied, or cheated them? They said yes. I ask, "what makes you think she isn't now?"

Not long after this event I come to realize the ranch North of my grandfather's place was Dr. Jones' ranch.

Dr. Jones came to a funeral viewing of one of my family members after said event. The only people that were in the room were my aunt, myself, and Dr. Jones. He came to see if I'd remembered. I asked him if he needed a new ranch foreman.

He told me he was going to let me go, but now he has changed his mind. It was a great connection to unexplained events—even at my elementary school—but not the head of the snake.

A simple Orange County pervert doesn't have the capacity to tell people what I'm thinking and twist the already tainted intel. This doctor alone doesn't have the ability to turn one of my closest friends into a paranoid schizophrenic with delusions of infidelity and attempt to attack me and then cause him to attack his wife. They twisted and broke this family.

But, this Orange County doctor has the money and connection to ask for professional help in destroying his potential exposure. Help from people who clean up messes or help from people who are just into maintenance all the time. Maybe his lawyer could find someone to clean up this mess—for the right price. Maybe Dr. Jones' lawyer could even have young people keep watch over me and be rewarded.

If an Orange County doctor can afford it, who else in this network has paid? Who is this brotherhood or agency with paying clients? This brotherhood of college students that must rape at least once as initiation. To steal someone's consciousness because their own is inadequate. This brotherhood of silence.

My aunt asked what Dr. Jones said to me, I told her not much.

If you, the reader; know anything about said events—it's time to reveal. Bringing someone to the lord can make up for multitudes of sin. I know the network is massive, but what difference does it make if you die today or 40 years from now? Twenty years goes by fast, and you have lived with the truth hidden. It's time to break these sons of bitches and send them all to an inferno of pain.

XVIII

When I was a child at Mauriceville Elementary—around 1987 or so—unidentified adults came to my school to perform a so-called hearing test without my parents' or my consent.

It was set up in the library where at least all of the 3rd grade was being tested in shifts. You had to raise your hand if you heard the tone coinciding with the ear you heard the tone from (left or right).

First these adults played tones that were obviously sounds coming from the speaker of the headphones I was wearing. Then I wore another set of earphones that were not plugged into anything.

At first I didn't hear anything—then I did.

As a matter of fact I could hear the tones without the earphones that were not plugged into anything. I continued the test without any type of speaker or head gear—and could hear the tone.

To my back right was a little girl sitting alone with her back to me. She appeared to be younger than me with long brown hair and a dress sitting in a blue chair that was only big enough for a small child to sit in.

While performing these tests, the unidentified male adult turned to the little girl for confirmation of some sort. She could hear the tones or she could hear me. When I say "me," I mean she could hear my thoughts.

The unidentified male adult was pretty resigned to the idea that the frequency would make me crazy—if I lived through it. He even said it with a little chuckle.

Weeks after the hearing tone test—more unidentified adults came to perform more tests. It is clear to me now that this was no test—this was psychological conditioning.

Mostly poor and brown children were in this group. They screamed at us like a drill Sergeant—telling us they were going to injure us and aggress on our person and property.

They grabbed one kid by the hair and jerked him around until he cried, wet himself, and broke to subserviency. They broke most all of the children in this group.

The children that broke with the most emotion from these adult aggressors became flaming homosexuals. Some for a period of time, some still remain. Maybe they were already—or maybe they were psychologically conditioned.

One man grabbed me by the throat—and said he was going to hurt me. I gave him my best Coonass smile and told him, "My daddy is going to kill you for this and if he don't—I will."

At least three adults - one being a woman—
were very concerned about my reply, and I aim to
keep my promise.

I can remember three women teachers fighting
with these people—physically and verbally. One
teacher stormed off to the main office—and came
back defeated. Which tells me the authority over the
elementary was involved.

The last I remember is sinking into silence upon
command—hypnotized. Like sinking from a subject
and then changing subjects. These women were fighting for me and the other children and I love them for
it still.

From then on the Mauriceville school faculty
treated me differently. This group of abusers then
provided faculty with their own story that scared my
teacher and everyone else that heard the story.

We were being taught authoritarianism at this
elementary school, under strict control with harsh
punishments and verbal abuse. These abusers set
everyone against me and I was under constant silent
scrutiny.

One teacher paddled me because she suspected
that I might be gay, which I was not. Others told students they would never amount to anything because
of a messy desk.

People would talk to me as if they knew my secrets and thoughts—to attempt to make me believe
someone was watching me at all times—and the story

would always be wrong. That's how I knew where the stories were coming from.

Other times people thought they knew how to redirect my attention with different commands such as look over there and see this other thing happening or even complementing or boosting my ego, I would act as if they could.

Some would try to remind me of an event or people they thought would stir an emotion, I gave them nothing and catalogued who and what to memory.

The worst is when people—much to their dismay—took it upon themselves to weed out their perceived evil—me. At a young age I had to fight grown men and became very good at it.

Some attempted elaborate ways to commit murder but were not strong, fast, or smart enough. You read correctly—conspiracy to commit murder.

Others, who fancied themselves intelligent developed schemes—such as producing plays that had names and events retold to fit their agenda of enraging me to have me aggress their person and property—all feeble attempts. They found that their lives went mysteriously to shit—others faded into the background, but not forgotten.

I took counsel from no-one, nor corrected them. I remained silent and patient.

What could these abusers expect but eventual retaliation? I come from people that are ready to defend to whatever level necessary. I think they were expecting a more passive group.

Just prior to all the events at Mauriceville Elementary, two teenage twin boys named Ronny and Johnnie—who even to a child looked as if their souls were in tremendous pain—devised a plan to place me in their attic before their father arrived home.

I was unaware of the true nature of this action. They were obviously scared and tried to remain calm while trying to convince me that it was just a game.

They advised me to remain quiet at all cost or I would lose the game.

I stayed up there a while, and as soon as their father got home—he at least beat them. I couldn't see—but I did hear them gag a lot from something being aggressively jammed into their mouths or they were being choked and they cried profusely.

They ran out of the house and he screamed and threw things at them.

I had no idea at the time what had happened nor was such action in the repertoire of a third grader. It was beyond my comprehension.

I was aware that the game was over—so I stood on the attic door to open it, mainly because no one came to let me out.

When I did the father was stunned and shocked that I came from nowhere. He was having cognitive dissonance from surprise. I stared a hole in that bastard and walked around his house as I chose while he followed me in disbelief.

In one of the larger completely mirrored rooms was all black and white checker board tile and an altar on the stage in the corner.

It appeared to be a Satanic sex cult.

I never said a word while in his presence.

Finally, I walked outside and stood at the ditch to watch the tad poles swim. He tried to give me a $50 bill. I let it slowly slip out of my hand into the water. He retrieved it.

I told no one.

The next day five men in suits came to my parents' door claiming to be the Census Bureau. My father opened the door and like most little kids I was in the doorway as well.

"We are with the Census Bureau." I could see that the driver of the car was Ronny and Johnnie's father. "No you're not." I told him. The man asked my father, "Is he psychic?"

My father could see the lie in their face. He told them he would kill all of them if they ever came back or gave us any fuckin' trouble. They believed him and my father slammed the door in their face.

They were indeed not coming back face to face —and we had no idea to what extent they would go to protect their way of life.

These twin boys' father or stepfather was very connected to " pillars of the comunity "—restaurant owners, state representatives, sheriffs, senators, judges, police officers, business owners, high powered lawyers, radio disc jockeys, a local doctor, his ranch foreman and son to name a few. They had quite a hold on our comunity and more—much further than our little town.

Could these actions have been all to protect one, or are they separate and simultaneous?

Around this same time a little girl a few years younger than me was raped and murdered by a couple that lived on the back side of my neighborhood. These Devils stole her from Larry's Antique Mall and did horrible things to her. There were demons all around us.

While riding the bus home from school I congratulated another kid because his step-father was elected State Representative of Texas. I will never forget the look in this child's eyes. It was real pain. He said, "He is an evil, evil man and he does evil things to me and my mom. Don't tell anyone or he will kill my mom!"

I said "It can't be that bad—I mean—does he whip you too hard?" He tried to describe abuse, but his words were taken from him and replaced with nonsense. So instead—with a face full of tears he just

looked into my eyes and I could see. I didn't want to
—I had no choice and it hurt.

Time passed and I heard the same tone from the hearing test and saw children turn on each other. The tone coincided with erratic behavior.

On one particular event in middle school—I heard tones and about seven children attacked one child that was bigger and taller than all the others.

We were all friends, and from out of the blue, everyone turned on the large child, beating him. I have discussed this with most all involved. All remember and couldn't tell you why they attacked him. As adults the memory is scary and confusing to all.

The large child became a big mean MMA cop who has now become a father and has his wits about him and is a good man who understands the constitution and liberties. For a while he hurt people. The conditioning that we all received was to produce a type of person to fit the needs of authority. It is against universal law and the perpetrators are guilty of crimes against humanity and more.

Think of how many more police and veterans of war are suffering from the same conditioning. They go into a trance and can't stop. They see actions beyond reality. They defend their lives with full force from minuscule actions because of conditioning.

It's an effective way of causing social discord. Like when Exonmobil gives their employees the options to either be fired for whatever reason, or beg for money on the street corner and maybe you can keep your job—to sow more social discord. How badly do you want to work here?

There were multiple times where I heard the tone and children flipped out at parties with paranoia and ran into the woods. Drugs were not involved. We had to run them down and catch them before they hurt themselves. Other times I would hear the tone and these young people stripped completely naked and ran around hysterically doing things with their body parts.

I'm speaking of a complete loss of control. No child at this age could want this kind of attention. This was something they would rather forget and can't. One more hurdle to keep your mind occupied.

The tone can cause you to talk louder, almost screaming, because you're temporarily deaf. You don't realize what level you're speaking and most people don't realize the tone. The tone may be causing people to actually be deaf. There were times when I couldn't hear at all, then I would spend time away from school and my hearing would get better.

One of the most sinister events was in the elementary cafeteria where a group that was disguised as a magic show hypnotized most all the children witnessing a man and woman on stage.

Two children—a boy and a girl—were on stage holding a glass Coca-Cola bottle inside a cardboard paper towel roll. The man was going to make the bottle disappear. He told all the children to think and say what he told them if they wanted the magic to work.

A few moments later there were children suspended in animation while others wallowed on the ground—sort of convulsing. It was clear to me that this was no magic show and was permitted not by Universal law. I got up and walked out—they followed. I walked around without any consequence to the playground to other classes and no one gave me trouble.

The show was over and the children exited the cafeteria like nothing had happened.

A few days later—early one school morning—the early kids could sit in the cafeteria and watch cartoons.

Rikki-tikki-tavi was playing on the tube in the front corner of the cafeteria and something triggered the children. They were acting erratic to a point where the male coach was very concerned and befuddled.

He knew something was wrong, but couldn't identify the cause. I knew but did not have the ability to speak. The words had somehow been taken from me. I had lost the ability to express anything germane to the subject matter.

I attribute my inability to psychological conditioning.

The male child on stage had severe psychological trouble and was falsely labeled as deviant for the rest of his school days. He chopped down a telephone pole to stop the noise coming from it.

Wrong pole—right direction.

He has adjusted and I know is a good person. We spoke about these events—and I assured him I wasn't going to let them get away with this abuse.

While speaking to him of these abuses I still had a conditioning left in my psyche. The other person in this conversation was a victim of these cruelties as well.

I verbally abused him in his own home. It hurt him and I'm sorry. I didn't know that it even happened until a few years later. This is how deep this conditioning goes.

I treated him like Americans would treat Mexicans in the 1980's—just a complete disregard of their existence and I was somehow more valuable than him.

He even tried to tell me his grandmother died—unusually I cared not. I remember the other person in the conversation saying, "Calm down. He's still in it." Referring to the psychological conditioning.

The entire high school football team—which he was a part of—turned on him with constant verbal and physical attacks. Even coaches aggressed on his person and property.

Coaches took him into their office and for hours told him how ugly and worthless he was and that they were going to aggress on his person and property. Not even the principals or any school leadership intervened.

Did he also hear the tones without the use of headphones? Most of the aggression from the coaches stemmed from a Dr. John song he liked as a kid.

Coaches were too ignorant to understand the song anyway. Most of the coaches were terrorizing teachers for passing grades to ensure good players stayed on the team anyway. All this for high school football? It's absurd.

He, I, and others somehow induced fear into whomever wanted control over the students and future employees.

A more noteworthy memory about this individual is one particular day before practice—while doing calisthenics—the entire team was chanting some crazy words directed at him. He cried and told everyone they were mean and God will punish them for their actions—he'd be right.

People remember and can't tell you why they did it or where the words simultaneously came from.

"A little food for thot"—first the tone then the action—in sort of a Handmaid's Tale way.

They even tried gaslighting me to ensure everyone—including me—believed their story. When I fly too high—these abusers aggress. I am favored by the lord from the beginning and learned to be directed as fast as a moment. I'm sure they are still wondering how I leveraged them so.

Gaslight - to manipulate (someone) by psychological means into questioning their own sanity.

Is there is a deeply psychologically imbedded—well connected—technologically advanced—Satanic sex cult in 1980s small town U.S.A.? Definitely all over Texas and Louisiana.

Then, it was unfathomable—now we have Jeffery Epstein, the ex-school teacher turned sex slave trafficker—the Boy Scouts of America, now bankrupt from having to pay lawsuits over sexual abuse—the Catholic Church and pedophile priests not yet bankrupt but still paying—and tech companies reading our thought, then turning wants into needs. It knows no boundaries and has been going on for generations.

Only a select few people know the actions of mind control for manipulation or have the equipment to transmit or disrupt brain wave activity. And they are able to have sex or control most people they encounter—mostly because the victims were broke like a horse early and are now easily identifiable prey. And no-one can understand how they accomplished this.

This is only a piece of the puzzle.

It's a part of the secret society that would die before they speak. The best way to retain members is to use ancient and new Edward Bernays style of mind control tactics to slip in their subconscious mind and force people to perform actions they wouldn't normally do.

Most of the time these people abuse their own families. After—if they could live with themselves after awakening from their psychological conditioning—they would have to become one of the aggressors to protect the secret. The abusers are victims as well—they have psychological trauma themselves and are terrified they will abuse their kids again. They just can't understand how it could possibly happen to them.

Some victims were conditioned to perform homosexual acts and when they realized what had happened they couldn't live with themselves. They either committed suicide or just block it out, causing more psychological problems. They were hypnotized like when a magician makes someone act like a chicken on stage.

After the act is complete, the aggressors have leverage over their victims. Some victims can now be placed in seats of authority like police, judges, city council, mayor, congress, senate, and even presidency.

While the mind controlled victim is in position of authority, the aggressors have other mind controlled victims remind these pawns of their actions to induce fear of exposure—causing the person of authority to do as they are told. The person expressing the secrets have no idea what they are saying, but expresses it anyway—expressing it by induced subconscious control.

These are effective ways to destroy a free society from within.

Ancient writings say that you speak in words without voice to those who dwell down below, and that man is complex and being of earth and fire. Wisdom is hidden in the darkness, and we must be the Sun of the Light without form, lit by the flame of our souls. We must find the wisdom and be LIGHT-BORN. It's in the heart of the flame, and we must tear open the darkness and shine a LIGHT on the WAY.

The children of shadows are here because man called them to gain great power. There are men who believe in darkness and use dark magic, calling beings from the deep. They are formless of another vibration, existing unseen by the children of Earth-men.

Only through blood could they find form, only through man could they live in the world. A long time ago Masters drove them back from where they came. But some remained hidden in spaces and planes unknown to man, living as shadows but at times appearing to man.

When blood was offered they came to dwell among men. Only by sight did they appear to be men, and are serpent headed when the glamour was lifted. They appear to be men among men and crept into councils as chief priests, scribes, Pharisees, and Sanhedrins. Their arts of slaying rulers and taking the form of rulers over men has allowed them to destroy man and rule in his place. Only by magic and sound could their faces be discovered.

The Masters came to teach man the secret Word that only man can pronounce. The Word is LIFE and in the name of Yeshua all things will come.

These serpents live still resembling man, and walk among places where rites have been said. They can be controlled and bound while in the flesh. If you seek the shadow, evil will appear. Only a master of brightness will conquer the shadow of fear. Fear is the great obstacle, but by being a master of brightness will you make the shadow disappear.

Attempting to reach the Barriers of consciousness is dangerous. But the HOUNDS of the Barrier can only move through angles, they're not free from curved demotions. Only a circle will give you protection from the claws of the DWELLERS IN ANGLES. It is only through strange angles unknown to men do they devour man's soul. Men who dare to reach the Barrier may be held in bondage by these hounds beyond time, held until the cycle is complete when the consciousness leaves.

Make your body into a circle and lose the pursuers in the circle of time. Create the circle that knows not angles, create the form that from the form was formed. Be cautious not to move through angles, for the HOUNDS of the Barrier moves through angles and never through curves of space.

If you leave your body, seek ye to move not in angles but curves. If you hear the bay of the hounds ringing clear and bell like in your being, flee back to your body in circles and penetrate not until you do so.

When you enter your body, use the cross and circle combined, use your voice and utter the word and in the name shall you be free. The circle and cross reminds me of the Medicine Wheel. Seek not to break open the gate beyond, seek your own Light and make yourself ready to pass on the way. Seek to find the light on the way.

Some Masons wanted to call a truce with me. They explained that a Mason from up north came down and told them he could show them how to take power from the deep. They didn't like the idea of performing these rituals at members homes because something always got fucked up.

"Fucked up?" I asked.

The man telling me cared only for himself and played off the repercussions from calling powers from the dark as "side effects," like nausea. It wasn't minor—because it caused the people who performed the ritual to sexually abuse their children.

In other cases, it caused siblings to commit incest. This cult is obsessed with anal penetration of children for punishment and control. It's as simple as poking them through their underwear with their finger or worse. Their powers derive from gossip and lies to protect themselves and their paying clients. All involved try to hide their secret and to do good to make up for the trouble they caused. Bringing someone to the Lord can make up for multitudes of sin.

Evil only cares about themselves, the only way you can trust someone who only cares about themselves is through blackmail. The fear of exposure and the loss of their own power is significant enough to sell or destroy their own family. They can get new families if they want.

They claimed there was a den of evil in a neighborhood they were getting rid of—the ceremony they performed caused this den of evil. Darkness spilled out everywhere around this mirrored room, it caused children to become very dark. Like the government, they caused the problem and also tried to give the solution by harming people—deflection from their our evil.

The Mason from up north set a trap for their souls—he threw the corn in the hog trap and they ran in. He masques as the bearer of light, just as they do. I couldn't believe they would sell their souls for short-term power. What they didn't realize is they were stealing it from others—or they didn't care.

They were in fact worshipping the Devil. They were angered that someone like me was born with the power they wanted. They said they were sorry for any trouble they caused and wanted me to forget about it.

I skirted around and asked more questions, and he said, "how about you just shut the fuck up and forget about it? I can't figure out if you're very smart or very dumb." I told him, "I hear ya."

I did forget about it, but then I remembered—I like to keep my word. A lot of Masonic Lodges have the letters AF and AM written on the front of their buildings and reside near schools and daycares. Are they stealing consciousness from children?

Does AF stand for Alternative Frequencies and AM stand for Amplitude Modulation Transmission?

So what's with the tones?

XIX

I can now see that some of these aggressors most likely came from the cellular telecommunication companies with the help of the Federal Communications Commission (FCC), and the Department of Defense.

The men in these organizations are indeed apart of secret societies, and used young people to watch over me and attempt—many times—to commit murder. They were promised high paying telephone company jobs if they could deliver information or me. They didn't get as high in the corporate structure as they thought they would. They watched too many episodes of 21 Jump Street.

The second set of headphones were bluetooth wireless. The tone without the speakers or head gear were from a millimeter size wave and powerful frequency that cellular technology excretes, were at the time most people submit to the idea of Tinnitus being the cause. Soon with more (EMF) electromagnetic frequency exposure more people will also experience this ringing in the ear.

Now there is a great number of people that can hear this high pitch squeal—like a water faucet slowly running, a dog whistle, or a hearing aid that is turned up too high making it audible for the non-user to hear.

The developers of the cellular technologies were experimenting on children at Mauriceville Elementary and are guilty of aggressing on our person, property, and peace of mind. They knew in the 1980s this technology would hurt people and continued anyway. This is malicious intent, conspiracy to harm and kill.

The new spin on the tone is longterm Covid-19 symptoms. The 5G infrastructure started simultaneously with the world wide sickness. The CEO of Texas Roadhouse chain restaurants committed suicide because he couldn't live with the ringing noise and probably figured it was never going away. I understand how he felt, except I recognized the sound and knew I could get away from it by leaving the cellular reception area, and turning off my phone and WIFI. Remember, it's not who hears the ringing, it's when will they hear it—some people are more exposed than others.

Now there are millions of people in the United Kingdom who experience this ringing noise and are resigned to the idea that it's Covid-19 related reported BBC News.

More than 40% of the people experiencing the ringing sound said the lockdown exacerbated the symptoms. It exacerbated the sound because of over exposure to electromagnetic frequencies, or EMF, a powerful frequency emitting from the 5G communications towers, cell phones, and WIFI.

The BBC is a perfect mouth piece for this disinformation. Not to worry, they have an app that can help with it.

This isn't the first time industry visited my school.

This tone can and does cause people to be controlled by an unknown source—making them have constant anxiety, hyper aggression, irritability, sweaty hands and feet, nausea, irrational thoughts, muscle ceasing in the neck and back, legs stop performing well, fatigue, chronic migraine headaches, accelerated heart rate, vertigo, numbness or tingling, metallic taste in the mouth, flu like symptoms, loss of taste and smell, and Magnetophosphene.

Your lungs feel as if they are being compressed, and people complain of Brain Fog. Brain Fog!?! The fuck ya'll think this is, Joe vs. the Volcano?

It is now being reported that women are bleeding uncontrollably from menstrual cycles pre and post menopausal terms, and having a great amount of miscarriages, vaccinated and unvaccinated.

Magnetophosphene - noun, a flash of light (mostly blue or red) seen when the optic nerve is subjected to a strong charging magnetic field (such as an MRI). This charging field causes current within the retina or visual cortex resulting in illusions of light. In one series, 8 out of 1000 people having an MRI experienced flashing lights.

I experienced this times ten at a soccer field where I exercise. There are two soccer clubs that inhabit the fields, one is easy to enter—the other (the Chris Quinn Complex) or more directly the Dynamo

Dash Youth Academy (a Houston group) has a painful frequency to keep you off their turf, literally.

The cell tower is less than 500 feet away and is very powerful. It caused me to see a yellow blob with blue line around it in my vision. I was temporarily blind and pissed my pants some. I crawled to a friend who helped me in the car—finally it subsided.

Is this technology being used as a security system to keep whomever is identified as unwanted subconsciously away without having to actually speak the words "get away" or be identified as racist? Because it's more powerful around areas where minorities live. Or is it just being emitted from the near tower?

This technology is too powerful to be in the hands of anyone and I would hate to see this technology fry a little soccer kid's brain. I can not direct attention directly at the Dynamo Dash specifically, I doubt a soccer club has the money or connection to complete this task. However, the powerful energy is directly on their field and not others.

The developers and operators of this harmful technology must have their liberties arrested, or they must be killed.

I have even witnessed it to cause a great number of people in Austin, Texas to have their necks and backs cease up simultaneously—and caused confusion. People couldn't get anything done or remember simple task—they walked around in circles. A lot of these people believe they need medication to focus—they need to be free from this dangerous frequency.

I have also witnessed the Magnetophosphene blue light flash disrupt analog signals on television. Not many people use "Rabbits Ears" antennas because they either have cable or stream their entertainment. As an experiment you should try to witness this yourself—especially if you experience Magnetophosphene.

When I experience Magnetophosphene, ambulances pick up old people in and around my neighborhood with stroke symptoms.

What was keeping everyone from sleeping in 2018? Why couldn't anyone sleep? Could it be technologies designed to take from our beings and cause, little by little, more pain and anxiety for control?

A high-pitched squeal kept me awake for a long time. I finally understood I could leave the sound by leaving the city limits of Beaumont, Texas. This is a clear indication that it's not traditional tinnitus. The boundaries of this frequency are in the same locations indicating an external source, cell towers.

I know this to be true because I visited several Ear, Nose, and Throat specialists for treatment and received Audiology tests. Some indicated that my hearing functions were normal, some indicated my hearing was exceptional.

My Honeywell Electric Smart Utility Meter-AMI (Type: A4RES) I received from Entergy electrical company that Alt tests and indicates it's recording every 5 seconds is producing a piercing tone and is making a cellular call that is 20,000 times too strong according to the permissible limit per 8 hour day which is 0.01 milliwatts per meter squared (mW/m2).

With great hassle, Entergy electrical company changed my meter for another. The piercing noise became stronger.

I realized that it doesn't matter what meter you posses and all the meters in the area are making a high powered cellular call every 5-10 seconds simultaneously to Baton Rouge, New Orleans, or outer space.

Wherever it's going, the power is harming everyone in a condensed housing area. All houses have smart meters and all are 20,000 times too strong.

My Trifield EMF meter model TF2 is peaked out, meaning it's the highest it can read. So the reading from all Smart Utility Meter indicates more that 20 thousand times above the permissible limit per 8 hour day—and we receive it 24 hours a day.

These electromagnetic spectrums charged by modern devices of conveniences have a direct impact on brain function and can induce visions by altering the electromagnetic field of the brain.

The communications companies could have picked any other frequency in the wide spectrum other than 2.4 GHz, which is the most destructive to our cells. But with the destruction of the blood brain barrier with metallic compounds emitted into our air, it is the perfect way to destroy and control humans.

What is the source?

In January of 2020 the 5G wireless network made a quantum leap in power and I hear the same tone while in the same room as a router, and all cell towers are at full capacity. I know because I can turn off and on my wifi router as a test to witness this intrusion. It's exponentially stronger when the wifi router is on. I only turn it on when I need it—like tap water.

The mysterious black towers that were being erected all over America in 2012 through 2018 were the beginning of the 5G infrastructure. Everyone was very concerned and after a short while forgot.

It was producing frequencies even in 2014 as experimentation on humans without consent and they aggressed on our person and property. It was the early release before the announcement.

On April 19, 2020, the Chairmen of the FCC (Federal Communications Commission) openly admitted on the television show Full Measure that 5G infrastructure went into full operation just prior to the COVID-19 virus.

In 1918 we had a pandemic called the Spanish flu. In the fall of 1917, radio waves were introduced around the world—causing biological systems exposed to the new electromagnetic fields to become sick, die, and the rest were suspended in animation which caused them to live longer and sicker lives— much like the inoculated in present day.

The Boston Health Department investigated the contagiousness of this virus—attempting to inoculate healthy people with mucus from people with this mysterious flu. Not one healthy person became sick. They even tried it on horses and were unsuccessful.

How did the Spanish flu travel from Kansas to South Africa in two weeks when the only mode of transportation at the time was horseback and boat? The entire world received the same symptoms at the same time. Much like today in the Amazon, tribes that have no contact to the outside world have become sick as well. Not to mention—the Industrial Revolution was in full force.

We had a pandemic during WWII. The military introduced radar—blanketing the earth with the energy of the HAARP program which emitted radioactive frequencies, which was later identified and released to the public.

In 1968 we had the Hong Kong flu and AIDS. Introducing satellites to the Van Allen Belt disbursed radioactive frequencies throughout the Earth again. Within 6 months a new viral pandemic began. We may be able to see Israel deal with a new autoimmune disease in real time from excessive vaccinations and powerful frequencies.

The first city to be completely blanketed by 5G frequencies is Wuhan, China. China also has a severe amount of air pollution making it impossible to see more than fifty yards. This is the very center point of the Corona/Covid virus. The communists used 5G on its citizenry as a stunning weapon after Hong Kong diminished their authority. They barricaded people in their homes because citizens were ready to attack the Communist Party. Hong Kong gave them hope, and hope is what the communists fear.

The permissible threshold limits per 8 hr day from this wireless network is 0.01 mW/m2 (milliwatts per meter squared). We are receiving thousands of times more power than 0.01 mW/m2 24 hours a day - 7 days a week.

The reading from my Trifield EMF meter model TF2 read 5.89 mW/m2 - 20 mW/m2 (milliwatts per meter squared) and beyond. That's 5 - 20 thousand times higher than the permissible limit per 8 hour day. Most of the time it just reads peak^ - meaning the highest it can read.

The lightwaves our data is carried on are becoming smaller and more compressed, carting more data, with more hertz or powerful frequencies, with more compressed waves or oscillations. The smaller the waves with more powerful hertz, the more data it can carry, it's a 5G selling point.

AM (Amplitude Modulation) transmission is producing a massive amount of invisible light, but are big waves, hundreds of yards tall and can travel far but can't carry text, video, or pictures. They can go through people and cities with less harm, other than the Spanish Flu.

The more common is FM (Frequency Modulation) and are around 10 feet tall and become more painful to the biology of all living beings.

Bigger wavelength and lower hertz bearing less information is claimed to be non-ionizing meaning it does not destroy, oust, or remove electrons from atoms. All wavelengths are ionizing.

Smaller wavelengths (such as millimeter sized waves from 5G that is the width of a penny) with higher frequencies is ionizing.

Ionizing means it can remove or kill an electron from an atom, just like a free radical that causes cancer in your body.

It's size and compression is small and powerful enough to become a laser beam and is used as a laser beam in Diathermy.

Diathermy - greek meaning literally to "heat through." It is electrically induced heat or the use of high-frequency electromagnetic currents as a form of physical therapy and in surgical procedures.

Diathermy uses three different techniques. Ultrasound, short-wave radio frequencies in the range 1–100 MHz (*shortwave diathermy*) or microwaves, typically in the 915 MHz or 2.45 GHz bands (*microwave diathermy*), the methods differing capability to penetration—it causes physical effects and elicits a spectrum of physiological responses.

It is used in surgical procedures to create higher tissue temperatures to destroy neoplasms (cancer and tumors), warts, and infected tissues; this is called Hyperthermia treatment.

In surgery, diathermy is used to cauterize blood vessels to prevent excessive bleeding. The technique is particularly valuable in neurosurgery and surgery of the eye.

Short-wave radio frequency is the high-frequency waves that travel through the body tissues between the coils and are converted into heat.

Microwave diathermy is radio waves which are higher in frequency and shorter in wavelength than the short-wave radio frequency.

Microwaves, which are also used in radar, have a frequency above 300 MHz and a wavelength less than one meter, a little more than three feet and smaller than most humans. Powerful and condensed small frequencies are harmful to all living beings.

Surgical diathermy is known as "electrosurgery" and is also referred to occasionally as "electrocautery." Electrosurgery and surgical diathermy involve the use of high-frequency AC electric current in surgery as either a cutting modality, or else to cauterize small blood vessels to stop bleeding.

Medical diathermy devices were used to cause interference to German radio beams used for targeting night time bombing raids in World War II during the Battle of the Beams which involved radar jamming and deception.

Radar jamming and deception - a form of electronic countermeasures that intentionally sends out radio frequency signals to interfere with the operation of radar by saturating its receiver with noise or false information. Sounds like what we are experiencing. Are they jamming our brain function with noise and false information?

A Midband spectrum is FM and can't get pictures, text, or video, nor can you perform surgery with it.

Cellular technology is more powerful than Midband. Your smart phone, television, smart meter, cell towers or 5G cell towers are all feeding from each other and become more powerful sending more information. They are together like mini cell towers. Together or separate they are ionizing and harmful.

Reminds me of Julius Caesar whom was killed by many. Brutus gave the final blow, but most all of the others couldn't be identified.

Science, the new religion, is easily paid off because millimeter size wavelength with powerful hertz are indeed ionizing, yet the FCC only cares about money, commerce, and the bureaucratic parasites that run the organization.

The smaller and more compressed the light waves are, the more harmful they become. Infrared or ultraviolet light is on the end of the spectrum and can't be seen by human eyes because our eyes aren't big enough to have the curvature to see. If they were, we would be able to see light coming from cell towers and all cellular devices as powerful as the Sun.

Ultraviolet light is ionization from the Sun, and there is naturally occurring ionization. Radiation is the movement of an amount of matter or energy.

However radio frequency radiation, especially at microwave frequencies, can transfer energy to water molecules. High levels of microwave energy will generate heat in water-rich materials such as most foods. This efficient absorption of microwave energy via water molecules results in rapid heating throughout an object, thus allowing food to be cooked more quickly in a microwave oven than in a conventional oven. Is this one of the many ways the earth will be destroyed by fire?

The water molecules in your body are cooked the same, also causing the water molecules to be deconstructed and evaporate causing heart complications, brain dysfunction, and many other dysfunctions related to dehydration. It's always relating to the theft and destruction of our water. Our bodies are made up of mostly water.

Radio and television broadcasting, cellular telephones, personal communications services (PCS), pagers, cordless telephones, business radio, radio communications for police and fire departments, amateur radio, microwave point-to-point links and satellite communications are just a few of the many telecommunications applications of Radio Frequency energy that are harmful to all living beings.

Electromagnetic radiation is waves of electric and magnetic energy moving together through space at the speed of light. All forms of electromagnetic energy are referred to as a "spectrum."

Radio and microwaves emitting from transmitting antennas are forms of electromagnetic energy.

They are collectively referred to as "radiofrequency" or "RF" energy or radiation. "Electromagnetic field" or "radiofrequency field" are used to indicate the presence of electromagnetic or RF energy.

The RF waves emitting from an antenna are generated by the movement of electrical charges in the antenna.

Power density is defined as power flow per unit volume, we measure this on a Trifield meter as milliwatts per square meter (W/m2) and is used to express intensity of exposure.

The quantity used to measure the rate at which RF energy is actually absorbed in a body is called the "Specific Absorption Rate" or "SAR." It is usually expressed in units of watts per kilogram (W/kg) or milliwatts per gram (mW/g).

You heard correctly, the SARS so-called virus is the adsorption rate of radio frequencies. The Corona/Covid virus and most other sicknesses are from radio frequencies, like the Spanish Flu. The branding

"Spanish Flu" was psychological conditioning to give Americans a feeling of disgust towards Spanish people—a better name would be Industrial Flu.

If you have a Smart Phone, go to Setting, General, Legal & Regulatory, then RF Exposure—this will give you an understanding but less accurate explanation of SARs exposure. Not all phones are the same, but the legal notification is there.

The RF caused your electrons to be ousted, the water molecules in your body to heat up and evaporate, and your cells release debris such as DNA, RNA, and proteins in attempt to purify itself (like cell vomit). This release of debris is what scientists will try to convince you is the virus. It's like saying your headache made you sick, instead of being a symptom of being sick.

Ungrounded children and adults have a higher absorption rate than grounded children and adults—ungrounded means humans who never allow their bare feet or body touch the earth. Two areas of the body, the eyes and the testes, are particularly vulnerable to RF heating because of the relative lack of available blood flow to dissipate the excess heat load.

The U.S. Government researched the biological effects of RF energy. The majority of this work was initiated by the Department of Defense, due to the extensive military interest in using RF equipment such as radar and other relatively high-powered radio transmitters for routine military operations. The Active Denial weapon uses millimeter sized waves with powerful energy, and can target your nervous system.

In the United States, some local and state jurisdictions have also enacted rules and regulations pertaining to human exposure to RF energy.

However, the Telecommunications Act of 1996 contained provisions relating to federal jurisdiction to regulate human exposure to RF emissions from certain transmitting devices.

In particular, Section 704 of the Act states that, "No State or local government or instrumentality thereof may regulate the placement, construction, and modification of personal wireless service facilities on the basis of the environmental effects of radio frequency emissions to the extent that such facilities comply with the Commission's regulations concerning such emissions."

The Telecom Act of 1996 prohibits local governments from using health or safety concerns as a reason to block cellphone deployment, except for one major measure.

"No State or local government may prohibit any entity from providing interstate or intrastate telecommunications services. Limited, nondiscriminatory exceptions to this rule are allowed for certain State and local government activities related to public rights-of-way, <u>consumer protection,</u> and other similar issues."

Consumer protection is what is not being performed. We must apply extreme pressure on the FCC, state, and local governments to withdraw this dangerous frequency or else.

Companies are designing protective clothing in an attempted defense. Animals, birds, fish, insects, and plants can protect themselves with protective clothing. Maybe some of you more ignorant consumers would want lead based paint to make a come back?

We have a dramatic quantum leap in the electrification of the earth that is unprecedented.

Read "The Invisible Rainbow" by Arthur Firstenburg. It chronicles the step by step electrification of Earth—within 6 months a new flu pandemic spreads over the whole world.

Viruses are not alive and have no cellular structure, and can not reproduce on their own—therefore can not be contagious. A dead thing and a dead thing doesn't produce an alive thing. It's completely opposite of reality.

A virus is actually an Exosome, an excretion of a toxic cell and are pieces of DNA and RNA with a few proteins (like cell vomit). This happens when the cells in your body become poisoned by our air, food, water, and the ionization from radioactive electromagnetic fields.

The poisonous air, food, water, and EMFs causes the body to produce the Exosomes or debris, causing cytokines to attack foreign enemies. Viruses are not the cause of sickness but an indication of being sick—cells release debris in an attempt to purify themselves, just as you would throw-up if you ate something bad.

This is one of the many ways science (the new religion) has collaborated with the pharmaceutical industry, and the evil men that lead them. It's an attempt to make you believe that humans get sick without cause.

The pharmaceutical industry is now receiving a pharmaceutical bailout with Corona/Covid virus tests, vaccines, and pill—the CDC and the pharmaceutical companies have become rich again. Tax payers of the United States will pay for every "jab" or vaccine worldwide and now there is a pill at $500 for thirty pills. There are 180,000 courses of Covid treatments (pill alone) allocated to the United States with more to come, reports PBS News Hour on December 22, 2021. That's 90 million dollars, not counting pervious test and vaccines with more to come. This will be the largest bailout to date—which is really extortion induced by fear.

In the 1980s, Anthony Fauci convinced everyone they had AIDS using Nobel Prize winner Kary B. Mullis's Polymerase Chain Reaction test or PCR method of analysis which is not designed to test for a virus.

"In the PCR, if you do it well, you can find most anything in anybody, it starts making you believe in the sort of Buddhist notion that everything is contained in everything else. Because if you can amplify one single molecule up to something you can really measure, which PCR can do, then there's just very few molecules that you don't have at least one single one of them in your body" says Kary Mullis.

Meaning you can find anything you're looking for depending on how high you turn it up. The CDC created a flood gate of false positives.

Asymptomatic is a lie. Fauci (the head of the NIH) got away with the lie thirty years ago. Fauci made a name for himself by pushing for higher doses of the deadly drug AZT, an old cancer chemo therapy drug too dangerous for approval.

Kary B. Mullis was hired to sample HIV in people's blood samples using the PCR and was told HIV was the probable cause of AIDS (Auto Immune Disorder) and found no proof. Lupus is an Auto immune Disorder. Fauci made it up because the CDC was losing money. The AIDS scare brought their finances back in the black and the men at the highest levels are in on the scam even today.

ABC's Nightline approached Kary B. Mullis about doing a documentary about his work; Mullis tried to convince them to cover the HIV debate.

In a July 1994 interview with writer Celia Farber of Spin Magazine, Kary Mullis expressed how he wanted to expose Anthony Fauci and Robert C. Gallo. Mullis said he'd be willing to chase the little bastard from his car to his office to ask a few simple questions of someone who spent twenty-two billion dollars and killed more than one hundred thousand people. Mullis said it had to be on TV, and that he was not unwilling to do something like that.

You can have a better understanding of this by reading Robert Willner's book "Deadly deception." Anthony Fauci, Robert C. Gallo, and all accomplices must have their liberties arrested immediately for crimes against humanity, genocide of gays, and mass murder.

On August 7, 2019 Kary B. Mullins age 74 died of pneumonia, three months before the shame of Covid-19. The person who could destroy the beast was eliminated before the initiation of the plan that continues time after time. They tried this with the Bird Flu, Swine Flu, Zika, and Ebola. They didn't have the mind control of Facebook to help before, and they will continue to force compulsory vaccinations until we kill them.

Why does the pharmaceutical industry need a bailout?

Johnson and Johnson is being sued unequivocally for causing ovarian cancer with their talcum power. The first few suits were in the billions.

Johnson and Johnson will now be able to pay their court tab with this new influx of commerce. Soon the babies that were exposed to this poison will be compensated as well. Johnson and Johnson also stopped vaccine trials for the Covid- 19 virus because their control groups mysteriously became sick—well no shit. Ever here of the Robert Wood Johnson Foundation? Robert Wood Johnson is one of the Johnson's in Johnson and Johnson.

Pfizer and Moderna have killed and made people sick with their vaccines. Johnson and Johnson was back in the lead to receive the original planned bailout, but their vaccines now give people blood clots in the brain.

All said vaccines are killing or causing sickness in their consumers. Injecting cellular debris or cell vomit into humans will cause people to become sick. Who'd have thought? Injecting humans with animal debris is beastiality. Our DNA isn't compatible with antibodies from cows or horses, vacca or vaccine.

Did anybody think to get an anti-bodies test after receiving a vaccine to see if it worked? Fuck no, you're to ignorant for that. It improves not on your immunity and is negative 7% efficacious—it's shown on the FDA's website.

Also the breakup of the opioid ring cut deep into the pharmaceutical industry's income. The pharmaceutical industry (like the auto industry) should have crashed and died from their transgressions. This is how real Capitalism works.

With the amount of money the government gave to the auto industry, they could have given everyone in America $650 thousand dollars. America would never go along with a pharmaceutical bailout. This is what the vaccine and test are for.

The Van Allen Belt captures and disperses cosmic waves for all the world—we are poisoning this protective region with hundreds of thousands of radioactive satellites.

Just as your phone emits the same radioactive frequencies—they are all water deconstructing devices. It is not compatible with health. The amount of metals you have in your body as well as the quality of water in your cells will also contribute to the electromagnetic field of your Specific Absorption Rate or SARS.

Vaccines have a massive amount of heavy metals—pharmaceutical companies claim it's to preserve the vaccine.

Also, Mexico, Italy, Spain, and Africa (all countries we share air and wind with—especially during the summer months)—produce a high amount of ash with heavy metallic compound combustion particles, and inorganic ions as a result of complex reactions of chemicals such as sulfur dioxide and nitrogen oxides, which are pollutants emitted from power plants, industries, and automobiles.

Heavy metals in your body make you more of an antenna for receiving and sending information. I know this to be true because I used a heavy metal cleans to diminish the ringing noise everyone is experiencing all over the world, induced by 5G.

Don't forget, Public Law 105-85 allows chemical and biological weapons testing on civilians without explicit consent.

NASA claims injecting vapor tracers in the upper atmosphere have greatly aided our understanding of our planet's near-space environment. NASA says Tri-methyl aluminum (TMA), Lithium, and Barium make visible the naturally occurring flows of ionized and neutral particles either by luminescing at distinct wavelengths in the visible and infrared part of the spectrum or by scattering sunlight.

Doctors who perform toxicology tests on patients find that all patients have a massive amount of Barium in their bodies.

In 1909 W.B. Clark of the New York times reported "cancer was practically unknown until the cowpox vaccination began to be introduced… I have seen 200 cases of cancer in vaccinated people and never saw a case in an unvaccinated person."

Edward Jenner is the inventor of the small pox vaccine, and gave his son brain damage with his vaccine. Vacca or cow is where vaccine comes from—to extract large anti-bodies from animals and inject them into human bodies—this is indeed Beastiality. Our species isn't designed to receive other species' DNA.

5G technologies indeed collect brainwaves and/or cause motor function interruptions. Brainwave interception can leave you gaslighted and concerned of how a select few can know what you think without you ever speaking a word of it. Sounds like science fiction, right?

Much like the DARPA Transcranial Helmets, the OpenBCI headset senses brainwaves and transmits them to hardware and software.

OpenBCI claims to work to harness the power of the open source movement to accelerate ethical innovation of human-computer interface technologies.

"We share an unfaltering passion for harnessing the electrical signals of the human brain and body to further understand and expand who we are."

Their OpenBCI system makes sense of an electroencephalograph (EEG) signal, a general measure of electrical activity in the brain captured via electrodes on the scalp.

The fundamental hardware component is a relatively new chip from Texas Instruments, which takes in analog data from up to eight electrodes and converts it to a digital signal.

Russomanno and Murphy used the chip and an Arduino board to create OpenBCI, which essentially amplifies the brain signal and sends it via Bluetooth to a computer for processing. "The big issue is getting the data off the chip and making it accessible" Murphy says. Once it's accessible, Murphy expects makers to build things he hasn't even imagined yet.

This is a polite way of releasing the information while trying to convince you it wasn't already being used or that the only way to obtain information is with electrodes connected to your scalp. Nothing could be further from the truth. As a matter of fact, not only can you send brain waves you can also receive information as well—beneficial or detrimental.

We are indeed electrical beings.

The city of Austin, Texas is collaborating with ID2020 on the MyPass project—a blockchain enabled digital identity platform. They claim it's for people experiencing homelessness.

Austin Blockchain Collective, Dell Medical School at the University of Texas, Austin, Travis County Emergency Medical Services, and the National Innovation Service is involved as well.

The MyPass project is funded by a grant from the Robert Wood Johnson Foundation—you hear about grants from the Robert Wood Johnson Foundation on NPR.

"We're building with, not for, people living the experience of homelessness and the service provider who engage with them" says City of Austin Chief Innovation Officer Kerry O' Conner.

Bloomberg Philanthropies is sponsoring a $100,000 grant and Mayor's contest that will award innovative cities—how far will $100,000 get Austin or is it for the Mayor? Blockchain is inextricably linked to the cryptocurrencies like bitcoin. You can buy a human hunt, children, or organs with bitcoins. Most hackers that hold information hostage demand bitcoin - kidnappers as well.

MyPass has been chipping homeless people around the world. "We are implementing a forward—looking approach to digital identity that gives individuals control over their own personal information, while still building off existing systems and programs," says Anir Chowdhury, policy advisor at a2i, the Bangladesh state's Access to Information Program.

While the ID2020 system mostly occurs in Third World countries, the team claims it's currently dealing with federal government officials in the United States to begin microchipping users with inoculations using the RFID chip injection system. Anybody getting inoculated?

ID2020 is additionally injecting refugees in war torn camps with its microchip shot via 2 debut pilot systems called "iRespond" and "Everest." What choice would they have—take the chip or go live your attrition.

The problem isn't homelessness or the past events of homeless people—its the inability to correctly diagnose the present. Because the medical equipment is so far behind, it requires additional data to ever come close to what the problem is—medically.

The people who want to chip our brothers and sisters on the street like livestock can only use their left analytical brain, and lack the capacity for real direct human to human compassion from the right brain. They are inadequate as humans and rely on artificial components to make up for their inadequacy.

The true problem is abuse at an early age, and a lifetime of bad decisions that would make anyone without proper guidance live a life of the ultimate fuck it.

Desperation from attrition can provide this ultimate fuck it as well—like refugees would experience. Refugees usually have two choices—go along or die. And yet we pat ourselves on the back for landing on Mars and humans on our planet suffer.

These people have let their egos and accomplishments override their better judgment. A massive amount of money and prestige can do that. Maybe they could flip the switch and have humans harm other humans by inducing imagery contrary to reality. It'll be better for our Masters to control their livestock's habits, wants, and desires—order out of chaos.

One way you can receive bitcoins is by data mining. What are they mining? They mine data on you and me—our habits, our wants, and our thoughts to better control human livestock for productivity.

Take away the feed a particular livestock likes until they do as the Master orders. Aggressing or limiting any human with these methods will result in the death of these aggressors.

Maybe they just want to shut down the motor function of the homeless—they can just go pick them up like sacks of feed and throw them away.

No one has the right to enter our consciousness. It's an invasion and they're guilty of aggressing on all humans' person, property, and peace of mind. The psychological invaders must be dealt with before we have to physically aggress on the invaders ourselves.

There must be legislation to keep governments, corporations, powers, and principalities from entering our minds, and using frequencies to harm person and property or there will be blood shed.

Refinery

As you read in Part 1 "Behind the Fence," I blew the whistle on Exxonmobil Downtown Beaumont, Texas Refinery. Exxonmobil went to great lengths to get rid of me from the site.

Exxonmobil had been gaslighting me for years—I remained silent. One particular morning—with thought alone, having not spoked a word to anyone since I woke up—I said over and over to my self "I'm going to break this canister through at Wharf 5, even if it isn't spent. I'm breaking it through anyway."

Meaning even if the carbon filtration system was unused, I was going to cause them to have to buy more by lying.

And so I set one of the many traps.

The first thing I do is go down to Wharf 5 (located on the Neches River), monitor the saturation level, and break the canister through. <u>I already knew it was really going to be used up.</u>

I follow proper procedure and go about my day. I check back later in the day and no-one has changed the filter. Exxonmobil only has an 8 hour window to rectify the problem, which is an ample amount of time. No-one attempted to change the filter.

I asked the unit operator why he hasn't changed the filter out. His reply was "it isn't saturated, and we're not changing it." He said that Exxonmobil had a third party (Darbonne LDAR Fugitive Emissions) check my readings and it was false.

I call Darbonne's Exonmobil representative and ask him about the issue. He said the canister wasn't used.

"What you're doing is calling me a liar. Get that fucking tech and bring both of your asses down to the wharf. We are going to read it side by side. Be down there in 5 minutes."

The technician used a flame ionizing detection unit (FID which burns the monitored emissions for detection) and they are well known to blow the flame out when the concentration is too high and no reading is received. Which is why Exonmobil loves this monitor so.

We indeed monitor together. His flame was out —I called him on it. I made him restart his monitor and we read simultaneously. It was around 500 ppm —VOC's (parts per million - volatile organic compound which is found in crude oil and is highly carcinogenic) and 250 ppm Benzene, everyone could see.

"You're out of compliance." I drove off and reported to my Exxonmobil contact. My Exxonmobil representative told me he's going to make sure it gets reported to TCEQ. I told him it was his job and hung up. Nice guy—inherited a hornets' nest. Darbonnes can only provide what Exxonmobil requires or your company will be removed from the site.

This was one of the many times I gave them the "rope a dope."

A woman I worked with on community projects —who is married to an Exxonmobil engineer—also divulged information concerning collected information. She only spoke what was told to her in more of a comical way.

She wanted me to vote for a jerk-off from El Paso as an official for Texas. In a fleeting thought I said to my self, "I wonder if she will give me some for it…" and then laughed to myself. Meaning trading sex for a vote. I would never really consider it because she is married, her husband is nice, and I'm not attracted to her. Like I said—fleeting thought.

The next day I see her at lunch and she tells me she only promised her husband a six pack of beer for a vote—implying sex was too high of a cost. I just smile and go about my business.

These actions have been being performed since that day in the Mauriceville Elementary Library.

It carried no weight nor did I believe she had an understanding of what she was doing.

Most of the time the true culprits go to someone like her and have them express something supposedly personal and deeply secretive to get an emotional outburst from me. That tells me the real culprits wanted me to harm the messenger. They most likely wanted everyone involved harmed using me as the instrument.

This is a tool in an attempt to make me or people like me not trust anyone or have them harm the messenger. It's a failure and only drags them from the darkness into the light. They are indeed cowardly and lack the warrior spirit. They don't even have the testicular fortitude to carry out these actions themselves—pathetic.

Exxonmobil and many other large corporations psychologically conditioned their employees to believe they must perform whatever action asked, because everything including your family, is riding on it. You must do it for love and the greater good. Without Exxonmobil pay—death soon follows. Exxonmobil will try to make you an accomplice to keep performing their biddings—break away from their corruption and keep your soul.

It can even keep Exxonmobil employees very quiet, not silent; strategic for self preservation. They are gaslighted and brainwashed by the computerized training modules they receive everyday.

Contractors must receive this training as well through ISTC in Nederland, Texas and other sites. The training is easy—the only thing you must do is look at the screen and listen to what is being said through the head phones.

Much like the conditioning children my age received in middle school. We had a children's news show we watched every morning called Channel 1 News for Children. Channel 1 News caused children in my reading class to become non-communicable for more than twenty minutes. My teacher in fifth grade was scared and angry by these actions—she had no remedy.

It was also proven to me that another faculty member had the power to command this scared teacher from a distance, as if one could communicate telepathically to another. It appeared that it wasn't voluntary and the scared teacher was having mental difficulties from this interaction especially after being gaslighted. She wouldn't retire because she was close to understanding the mystery.

Channel 1 News was designed to keep future employees compliant. In elementary school we watched cartoons much like the Wizard of Oz, except this cartoon was tin man heavy. Mainly because he had to have oil to survive. I can remember little girls crying because the tin man was in pain because of the lack of oil.

Anywhere there is a refinery or industry children will be taught repression. Their hair must be acceptable in length and style, approved clothing and style especially on the Gulf Coast where most petro industry resides.

Exxonmobil gives employees the option to beg for money on the side of the road to feel what it's like without Exxonmobil. This option is given to employees who aren't compatible with the program or employees that aren't thinking the way Exxonmobil requires. Over all this action is designed to cause social discord throughout our society. Look at the homeless and remember, you could be next.

How many people have they driven to perform mass shootings by gaslighting them? It's not only large corporations that have access to this information and technologies. The police force has access to it as well.

Police

Back in about 1995 or 1996 a friend and myself left a Jerry Jeff Walker concert at the old Beaumont rodeo arena in the middle of the night—drunk. We drove to College Station were he claimed there where girls waiting on us.

We did not make it to College Station because we were arrested in Cut and Shoot, Texas. The officer that booked us interrogated me while in the presence of another official sitting at a table with him while my friend and I stood in front.

He asked me many questions. The one that stuck was when he asked me where the bodies were. I had no idea what this officer was talking about. Through a few small questions of my own, I came to understand they had a machine that reads brainwaves and it registered me as a deviant or a killer.

With his own words he told me this. He wore high boots with his pants tucked inside and a cowboy hat. He never caused me trouble other than making me spend the night in an over crowded drunk tank. He thought his machine could give him inside information—he was now the one confused.

This was one of the many times I encountered the police with the same story.

Another time I set my 1979 CJ-5 Jeep in front of my aunt's house to sell. A cop rolled up and wanted to know if I stole this Jeep. Jokingly I asked if he had a warrant. He became befuddled and I flipped the script. The end of the conversation was when his technology told him I was a criminal. I made him run the plates then I told him to get off the property before he gets his ass whooped. He did because of my truth and his dismay from his technological disappointment.

I had trouble with the police but hardly a record. I was targeted but never charged. I wonder how many people with learning disabilities admit to things they didn't do—just by giving them a little information about themselves and gaslighting them into believing, maybe they did?

This is some Devilish shit. I know people with high intelligence that suffer the same harassment I did. That tells me their equipment is flawed or wants to target people who think for themselves. This technology has ruined many lives and made murderers and liars out of good peace officers. They indeed are guilty of said charges and many more.

You can manipulate the police as well by expressing the correct emotion that is counter productive to fear. They are counting on you to give in when they give you a little truth, but they don't have the real story. Like Satan, he and they have half truths, which are not truths at all.

I have fed false intel to many demons who vomited it back to me and attempted to use it as leverage—it was ineffective. For a short while, when friends and family relaid false intel, it caused some discomfort. Only because of my desire and resistance to injure them permanently. They know not what they say.

Energetic Attacks

XX

Have you ever been driving along and feel all the energy plummet from your body and you're completely exhausted? Once you get a few miles down the road you feel better.

Some people think "must be an Indian burial ground" or "bad energy over at this place." It's more than just a feeling.

I was visiting family in San Antonio and wanted to visit the Guadalupe River State Park. They were not interested in going, but wanted to know what I thought of the park.

As I exited off Bulverde on to State Park Rd 31, images and feelings of severe depression was induced, followed by severe anxiety, and the tone. I know it was induced because it wasn't my own—I have felt this pain before.

I continued down the road in defiance and the pain grew stronger. I thought to myself—if this were a military action it wouldn't stop me. It would only make me want to kill everyone when I reach my destination.

I got to the end of the road and some bitch that worked the gate treated me badly. I told her I knew there's bad energy leading up to the park and I'm not staying. She knew what I was talking about—gave me a shitty look and opened the gate to exit.

I asked my family what they experienced—they told me pain and anxiety. Their poor little babies cried immediately and didn't stop until they were far away from that place.

The technology at the Guadalupe River State Park stole theirs, my, and other people's energy—caused severe anxiety and depression—leaving them completely drained and exhausted. I know people not related that experienced the same assault and we all want to know why and how this happens.

Why would they need to keep people away from this state park? To keep the over populated San Antonio away from destroying the eco-system? Or are they hiding something on this site that is important enough to harm humans?

There are many places in Texas with these harmful energies.

I understood I could buy used tools cheap at the State Surplus Barn on Bolm Rd and Hwy 183. The building also houses the Railroad Commission.

I looked around at a lot of office furniture but not many tools—mostly junk. I was sure the Surplus Barn employees got first pick for resale or use.

I found a 12" cast iron frying pan I liked and proceeded to check out—there were a lot of people in line with items for check out.

The State Surplus employees were fumbling around and were talking to each other about a program they weren't sure of and were nervous about this technology they were having trouble with.

They got on the phone with tech support and spoke with few words as if they didn't want the people checking out to know the information they were receiving. The State Surplus employees fixed the so called "problem" and I could feel the energy start—causing anxiety to everyone in the room. It was as if the air became stale—it felt like our lungs became restricted, and I could hear the tone.

The room and line was full of people ready to check out. In less than a minute everyone had abandoned their found items, leaving pissed off. Not even the person at the front of the line stayed to check out —just me.

I could feel it—but it wasn't near enough to throw me from a simple checkout line. I smiled and asked a few smart ass questions about what they have done. They were tight lipped and big eyed.

Was this action designed to keep people away from surplus goods? Or psychological warfare to keep people away sub-consciously. Why would they need to keep people away? Some of the people that have experienced these energies at the State Surplus Barn on Bolm Rd and Hwy 183 are afraid it's black magic.

The third law of science fiction writer Arthur C. Clark is "Any sufficiently advanced technology is indistinguishable from magic." People that have never experienced these feelings would feel like this technology is black magic—especially when it's harming them.

All of the cell towers in Beaumont, Texas are at dangerous levels of power especially on College Street, and on Major Road. It appears to be strong everywhere poor black and Mexican people live. In this case College Street is Highway 90 and leads to our military port.

Is this a way to clear a path for an invading force if there is a confrontation? College Street is abandoned like Port Arthur and Port Arthur has the same frequencies. These frequencies have the ability to control commerce subconsciously.

The telecommunications companies on behalf of other companies can kill businesses and corral like cattle consumers to other more mainstream corporate business. They can drive people away from the blast zone from Exxonmobil Refinery, it's only a matter of time before this turd of a plant explodes. They can clear a path for the military or of a shadow government.

You can feel the limitations of the energy field when you leave Jefferson County on Hwy 105 and on I-10 in Rose City.

The Federal Court House in downtown Austin has the same energy. The courthouse employees made a big deal about my sombrero I wore while attaining a new passport—most courthouse employees being Latino. Really they're just a bunch of city people and have no idea what a hat is for—to keep the Texas heat off! I believe they used this technology in retaliation of what they perceived as racist.

The lady that completed my passport told me I should go to a place where she claimed she played basketball—I saw the place. It was to have me go to a place where more city Latinos would be angry at my hat and retaliate in violence. The only thing that would have happened is I would have shot and killed many aggressors and would be wrapped in legal hell —ignorance is rampant.

Most of these people don't have to be in the sun and most of their lives inhabit conditioned air. Real Mexican laborers want to know where I purchased the sombrero and how much it costs. I supply the information because it's useful.

How did we get to a point where city Mexicans don't know what a hat is for or who the Texas Tornados are? Or the fact that immigrants are employed and are now in a position to provide immigration status; looks like liberal fuckery to me. The fox is watching over the hen house. More importantly, employees of the State of Texas have the ability to use non-lethal measures to harm anyone who is not in their society as they see fit.

If these and more state properties have devices of non-lethal destruction then it's obvious that the governor of Texas indeed knows of the said device and is using them as he sees fit.

Greg Abbott, a Republican, is being sued by the Republican party for spending more then $295 million on a deal with Frisco-based MTX Group to perform Contact Tracing for the Corona/Covid virus. The invasion of privacy isn't the only function for contact tracing.

Through Contact Tracing they can punish you for speaking against Greg Abbott or the New York Times, or anything undesirable. If you speak with anger on social media like Facebook or Instagram there is a dramatic elevation if frequency that is painful to the nervous system. It's like an invisible shock collar designed to socially engineer you.

All the painful EMF attacks I received from the telecommunications companies on behalf of state entities and social media companies like Instagram and Facebook caused me to have overexposure causing selective and destination tinnitus.

Bolivar Peninsula's cell frequency is so powerful it's not a ringing sound—it's a sizzling sound and causes blurred vision. It starts when you come into the vicinity of the Intracoastal Waterway. As I said in the Water section: "if they drive us away from the coast, we are lost."

Greg Abbott is an infringer on privacy and is well connected with the communist party.

I cussed Greg Abbott on social media calling him the crippled communist. A few hours later the ringing noise I have experienced since March of 2020 became powerful and I became weak in the knees.

Through Contact Tracing, the Social Media Gods, like Instagram or Facebook, can attack anyone they choose or view as an enemy. This is too much power for anyone to possess. Absolute power corrupts absolutely.

The Active Denial System - <u>military.gov</u>—is a non-lethal weapon that can impact a single human—even in a mass crowd. It can stun you, compress your lungs, and give you flu like symptoms. It uses cellular frequencies and sound waves to incapacitate their target. They have expanded this technology into 5G, the Trojan Horse.

When I intentionally say something hateful on social media or anything unpopular like celebrating the destruction of the New York Times that was proven to be untrustworthy by Federal Court, or the Satanic acts of Israel against the Palestinians, or the pinpoint locating of severe pollution, they retaliate with a painful high pitch frequency that starts and lasts for days.

The tech companies, media, and social media want to socially engineer us—they want to become our God. There may be someone acting on their behalf, perhaps the contractor who provides this service.

They don't need the Active Denial System now because they have the 5G Network to control you and perform the same functions.

The telecommunications companies are using this powerful, dangerous, millimeter size oscillation as a laser beam. The 5G has no more of a wave than a bullet, any wave smaller than 500 yards is detrimental to all living beings. It could be that all unnatural light waves are dangerous, or at least an over exposure with a great number in a condensed area is detrimental to life and health.

Animals, plants, insects, and the Earth can't wear protective clothing to shield them from the deadly ray, nor do they have the cognitive ability to do so. My poor dogs cry and scratch at their ears when the frequency is high.

Cellular technology is out of date, and every human is using it. In attempt to continue business as usual they are hurting us or they are involved in the communist take over.

Greg Abbott can stand up for your gun right because guns are useless against this broadcasting weapon. They don't need an army, the elite have the ultimate weapon—a Death Star if you will.

Little by little these demons are weakening us. You can give not a millimeter to a demon. They will continually push to your protest, wait till you're calm, then push to your protest again. Before you know it, you're so far behind the line from which you started, you can see not the starting line anymore.

Read the book "Ordinary Men" by Christopher Robert Browning. You must be ready to kill or die to save your soul.

The Communist Party

XXI

On February 6, 2017, Chinese Consul General in Houston, Amb. Li Qiangmin and Madam Zeng Hongyan visited Austin and met with Texas Governor Greg Abbott, State Secretary Rolando Pablos, Austin Mayor Steve Adler and ten State Representatives separately. They shared views on cooperation between China and Texas in the fields of energy, agriculture, medical care, and cooperation between Chinese cities and Austin on creativity. Deputy Consul General Liu Hongmei and Commercial Counselor Zhou Zhencheng were present.

On October 19, 2017, Amb. Li Qiangmin, Chinese Consul General in Houston, attended the Grand Opening and Ribbon Cutting Ceremony of the General Chennault Flying Tiger Academy and delivered a speech. An academy of Communists.

You can find this on the Ministry of Foreign Affairs of the People's Republic of China's web page <u>fmprc.gov.cn</u>.

On December 18, 2019—just a month from the pandemic—Austin, Texas Mayor Steve Adler is seen on his Twitter account shaking hands with the mayor of Beijing and it reads, "With the Mayor of Beijing, discussing the new era of increased direct city-to-city interaction and cooperation."

It's authentic because there is a blue check by Mayor Adler's name. Mayor Adler wants to cooperate

with slave labor, state sponsored murder, and a police surveillance state. Mayor Steve Adler is in love with money, technology, and wants to be openly a part of the Chinese communist "Smart City." It is said that Mayor Adler has over $300 million in real-estate he would like to control better.

Greg Abbott's biggest connection to communism is the oil business in Odessa.

Like I said in the Water section—most of the oil and land in Midland-Odessa belongs to Chinese companies like Surge Energy and Yantai Xinchao to name a few.

So when he held the roundtable discussion in Odessa on January 27, 2021 at Cudd Energy Services to discuss his executive order protecting oil and gas in Texas, he must be talking about protecting Communist Chinese companies like Surge Energy and Yantai

Xinchao to name a few. Cudd is an energy service company, the owners of the oil and land are communists.

The main focus you need to have is the Chinese are not Chinese—they are Communists—really they are Authoritarians. These Authoritarians commune not, however we must identify them as they identify themselves.

Communists have an unlimited amount of money they can generate through the oil business, insurance companies, technology, state sponsored murder, and slave labor. The communist economy is designed to destroy Capitalism.

The Communists (Authoritarians) can turn Americans into billionaires if they help the communist party. Most billionaires you know of became billionaires from their connection to the communist party.

Watch the PBS News Hour seven part series on Communist China. Nick Schifrin reports as part of "China: Power and Prosperity," with support from the Pulitzer Center. It will explain most all series of events that lead up to what we are experiencing today. One of the most important segments is the ten minute episode called <u>How China's high-tech 'eyes' monitor behavior and dissent.</u>

The Communist are indeed facilitating a surveillance police state globally.

A company called Ping An Healthcare is the second largest insurance company in the world. Mon-

ey generated through this company helped build Ping An's technology company.

Ping An developed a new facial recognition software that judges weather someone is a good driver, and feeds data into Ping An's data base. Ping An can also use facial recognition to detect if loan applicants are lying about their identity by examining 90 different expressions. In America we have XR Vision to perform the same task.

This reminds me of Jordon Peele's video on deep fake. Where he made a video of presidents Obama and Bush saying the words he spoke and using the facial expressions and gestures Mr. Peele performed.

Ping An uses a social credit score based on data entered into users' phones, such as data collected from the users mobil phone bills, or shopping records and habits.

Ping An promotes the idea of wanting to work with people who have nothing to hide. But in communist China, who decides who has nothing to hide? The Communist party decides.

The Communist party has a computer program called Sharp Eyes. The five most surveillanced cities in the world are Chinese—with over 200 million cameras watching for insurrection.

This technology not only traces peoples faces, but how they walk and the sound of their voice. The cameras can judge behavior, the same as our police force who wear the same cameras on their bodies. These police body cameras perform the same function and are severely flawed.

Exxonmobil uses behavioral data that left them quite confused when they tried to understand me. These computers, or the data collected by personnel, are no match for the human mind. In fact, Exxonmobil is having a hard time with the sudden changes of algorithms from all their employees—whoops.

If someone jaywalks, that person is publicly shamed with their face broadcasted on public screens and their social credit score will be reduced by 2 or 3 points.

It's kinda like when everyone went after every suspected "Karen" with cellphone video for not wearing a mask or any other trivial action performed by free humans. We have become a nation of snitches.

Karen - a pejorative term for women seeming to be entitled or demanding beyond the scope of what is normal. The term also refers to memes depicting white women who use their privilege to demand their own way. A woman who wants to speak with a manager, an anti—vaccinator.

I'm a little ashamed to have to refer to such a term—however it depicts the times. Karens may be raging bitches, but at least they know how to have a backbone.

This credit score is so important to the Communists that it produces a national credit score magazine.

Why must there always be winners and losers? To have you believe you're in competition.

People with regular credit scores pay regular price, people with excellent scores pay only 80% of the regular cost. What the Communists consider good behavior is rewarded all across society and punishing so called bad behavior is enshrined in the social credit magazine.

What they will produce is a backlash from good. Repression leads to excessiveness.

Nick Schifrin asked the editor of the magazine if rewarding the people with good behavior and punishing people with bad behavior works well.

Her reply was, "Of course it works." However, the question made her very uncomfortable and so concerned that her and her staff exited the room not realizing they still had microphones on.

As they spoke in a separate room they said things like, "Don't talk about the government—talk about companies," "We need to be calm—we can not refuse to be interviewed," and "not too rigid or serious."

Ten minutes later she came back to finish the interview. Nick Schifrin asked if everything was ok. She replied that everything was ok.

People in Communist China, Exxonmobil, or American universities say and do what they need to for self preservation.

Self preservation can be as small as being able to listen to what music you enjoy, or as medium as preserving your job, or as large as saving your actual life.

Professor Jeff Rubinen who teaches Introduction to Information Technologies at Syracuse University enjoys controlling student behavior.

Syracuse University uses the SpotterEDU app to track students that connect their smartphone to the campus WiFi network. The University claims the app's purpose is to boost attendance and logs information into a campus database.

"They want those points," says Professor Jeff Rubinen to the Washington Post. "They know I'm watching and act on it. So, behaviorally, they change."

He was referring to the app's ability to socially engineer his students like god.

He is their new god and has control over their consciousness. SpotterEDU works with more than 40 schools in the United States.

Rick Carter, a former basketball coach, is the company chief executive officer of SpotterEDU. Carter developed the app in 2015 to watch over his athletes. The app uses bluetooth beacons that are placed on walls and ceilings to ping students smartphones. It notifies advisers of the students presence or deviation of habits. Sounds like Rick Carter would do well in the communist party.

"We have to make sure they're doing the right thing" claimed University of Missouri associate athletic director Tami Chievous. I guess Chievous is the authority on the "right thing."

Students are measured and analyzed through Degree Analytics, according to data scientist Aaron Benz.

Benz designed a set of algorithms to look for patterns in students' behavior and automatically flags when their habits change, called scaffolding. Students who deviate from their everyday rhythms are flagged for anomalies. I'd like to know Aaron Benz's habits.

University tracking information was reported by Washington Post.

What these Universities are really doing is ushering in communism and making compliant employees, subservience to control from authority, and fear of being punished at all times, even when you sleep.

These universities are in direct alignment with the Communist Party and must be punished for aggressing on freedom and liberties of Americans. We must squash this police surveillance state anywhere in America——by blunt force if necessary.

The education system has been infiltrated by the Communists generations back. Time has helped this treason feel normal. It is not normal—it's the constant abnormal.

Critics and myself claim the communist government can use this technology to punish anyone who opposes or criticizes the Communist party.

Hong Kong protesters claim mainland China is exporting a system of surveillance, and they indeed are, even to the American educational system.

Viper Networks is an international leader in the LED lighting products and integrated systems markets for Smart City projects. The Apollo Smart Lights provide municipalities with the ability to control and adjust street lighting while enabling them to monitor the streets, enhance security and manage traffic; all while significantly cutting costs, and reducing the environmental footprint while creating new revenue sources for both the private and public sectors.

Surveillance in every town in America. This is only one company participating in the snitch culture.

Communist China is trying to destroy Hong Kong and its aspirations of freedom and democracy. Freedom and democracy belong to Hong Kong and all lands around the world.

When you speak against the Communist party you are arrested and charged with subversion. Communists have kidnapped many citizen journalists and no-one has seen or heard from them since. Honk Kong went from the rule of law to being ruled by law, fear, and intimidation.

Corruption of authority leads to imprisonment of opponents of Xi Jinping, president of the People's Republic of China and the Communist party.

Protesters spray paint cameras and cover up their faces. They know the Communists can find them in minutes and they are scared.

Protesters are scared because surveillance goes from cameras to inside their phone. Most protest rallies are organized off line because police hack into their messaging apps. Surveillance is indeed omnipresent—-automatically and instantaneously.

The protest at the capital in the District of Columbia is a great example of contact tracing and facial recognition technology from a company called XR Vision.

The protest wasn't that violent—only a few people got killed and all the House and Senate escaped. If you feel this chicken shit protest was scary or appalling then your personal constitution is weak. You are the cause of America's weakness. If the mere sight of a weapon causes paralysis then you need to look at it closer.

Guns scare people because they have psychological conditioning that induces involuntary compulsions of suicide and suicidal thoughts. I have taken a poll for more that ten years, and 98% of the time, the person fearing the weapon has an impulse to kill or harm themselves. This is why they want to destroy all guns.

Residents of Hong Kong fight the Red Army with umbrellas, bow and arrows, and Molotov cocktails. They fight for freedom and their lives.

The D.C. crowd was misled by the church of Qanon—at least they're ready to stand against tyranny, even if they can't identify the source. Once we agree and pinpoint the source of oppression and it's true to all eyes—we will be unstoppable. The tree of liberty grows thirsty.

I do wish it wasn't a poor metro cop and veterans that had to die. The culprits of our internal decay will be seen by all.

WeChat is the country's most used communications app. If you send a message about something sensitive like the 1989 Tiananmen Square Protest the recipient never receives it. Controlling the mind is most important to Communists.

Telecommunication companies in America limit communications as well. In America, Facebook, Google, and Twitter identify their ideological enemies and shadow ban their information.

Facebook, Google, and Twitter now perform the duties they claim the government is lacking—censorship. All because they were personally offended by the president's private words—pussy.

Noam Chomsky writes "The smart way to keep people passive and obedient is to strictly limit the spectrum of acceptable opinion, but allow very lively debate within the spectrum...." Since the pandemic, Chomshy has revealed his true sentiments towards freedom—which is not very free at all.

Facebook, Google, and Twitter are now the authority on what is true or not. Facebook, Google, and Twitter are now the "Thought Police" that George Orwell tried to warn us about.

In America, you are free to think as you choose—even if it's racist or prescribed indoctrination. You're not permitted to aggress on someone's person or property even if those ideas are in direct alignment with said ideologies. You can think it, but don't do it.

On January 21, 2021, the New York Post reports, Twitter is being sued for telling child porn and sex traffick victim, John Doe, that images of him at 13 years old didn't violate their terms of service. I guess that fits their scope better than freedom of speech.

Twitter, along with Public Good Project and Ad Council, bribed micro-influencers and celebrities to promote Covid-19 vaccination and to promote the Democratic Party.

The AD Council raised $50 million from Bank of America, Facebook, GM, Google/Youtube, NBCUniversal/Comcast, Salesforce, Verizon, Walgreens, Walmart, JPMorgan Chase, and Target.

Facebook deleted a 120,000 member group who posted stories of vaccine reactions. Straying from the Covid-19 narrative isn't permitted, whatever Facebook tells you is permitted is permitted.

Facebook helped infiltrate personal information from its users to build the surveillance police state we now live in. Facebook has a system that is able to freeze commenting on threads in cases where their systems are detecting threads of hate speech or violence in the comments. Facebook causes every user a little more pain everyday—and the users can't put the needle down.

Freedom of speech is painful—especially when you don't agree. The real problem is Facebook has run its course because it has built its castle on shifting sand and has been infiltrated by darkness. Because of its lack of integrity to hold back the water, it will soon disappear—Meta or not.

Facebook doesn't want the party to end and would do anything to keep the massive amount of money rolling in, hence the rebranding as Meta—all things end. I don't blame them for wanting to survive, it's just not possible because it's not true in the universe. If it were, it would last.

Without a library, dictionary, or encyclopedia, Google and Wikipedia supplies you with what truth they see fit. If you don't fit Wikipedia's Zionist narrative, as punishment they will turn on you.

Just ask Stefan Molyneaux, who debunked a white supremacist assembly in favor of true freedom, which is the right to assemble whatever your views are. Now Wikipedia has identified him as as far-right white nationalist, conspiracy theorist. There's really not many places to search for information.

Back before the internet, I knew a child that had an imaginary friend named Googles. Googles would talk to this child and give him information.

The child would hold his hand up by his face in kind of an okay sign and look through it like a looking glass. He would tell me you can look through it and ask Googles questions and it would give you answers.

I asked him if the looking glass was like what the peanut wore, referring to the Planters Peanut man who wore a monocle.

He said no, and with his hand he held a circle and then made a line with his finger pointing down—like a magnified glass.

What is this being that calls itself Google? And why or how could this being only talk to a child pre-internet and sometimes give him detrimental information on life. Google may not be just a computer generated information site.

Amazon removes books critiquing the Transgender movement. Amazon is so elevated it's now banning books? Apple products are completely filled with tracking and surveillance technology. Its best feature is to monitor every aspect of your life. All said technology companies work for the NSA and are collecting data on every user.

The whole problem with these companies and companies alike are the monopoly they posses. We must destroy 3/4 of their property, tangible assets, and cease 3/4 of their monies.

A monitory fine is no match for money generated by communists through slave labor and state sponsored murder—technology comes mostly from communist countries. Monopolies aren't permitted in America—for damn sure not allowed in Texas—we are reaping those seeds that were sown now.

It's probable because not many people are intelligent enough or have an understanding of Capitalism. They can't stand up to industry because of a lack of brain power and a backbone.

In Ecuador, whenever foreign media interviews someone who is a Communist critic the police show up at the interviewed person home using Huawei phones.

Huawei and ZTE are a $100 billion phone and technology giant and are the world's largest provider of telecommunications equipment. Both said companies are fronts for the Communist party.

Founder and CEO Ren Zhengfei is a member of the Chinese Communist Party and a former officer of the Chinese military. International researchers say Huawei collaborates with analysts from the Communist Army, or PLA.

Their gift is a Trojan horse for authoritarian governments to control and engineer everything in society. All of our technological conveniences are a Trojan horse—even your appliances.

Places like the Philippines or Ecuador who purchased "Smart Cities" from Huawei to fight the cartel are now feeling the pain of this police surveillance state. They also may not be able to continue payments on the smart city. Which mean Huawei or Communist China will own it soon.

Huawei or Communist China operates with or like the International Monitory Fund (IMF). Giving loans and stipulations the buyers will never be able pay—and they foreclose. All said parties are economical hitmen.

United States officials describe it as the symbol of Chinese hi-tech suppression and beholding to the Communist party.

President Trump mostly blocked U.S. companies from selling technology to Huawei, and we should do more—much more.

It has recently expanded it's 5G of Fifth Generation phone technology and has sold 150 thousand Playstations outside of China.

Back in the late 1990s, Huawei was responsible for thousands of Chinese children having seizures using their video games.

Ever heard of Polybius? It's a video game that appeared in arcades and became so popular that fights broke out over who would play next. Players began to have nightmares, insomnia, memory loss, headaches, and suicides.

Unidentified men visit the machine, collect data, and leave without speaking to anyone. The game vanished months later and hasn't been seen since. I know people who played the game and experience these symptoms to this day.

How about Grand Theft Auto? A game where you steal cars, run over people, beat up and kill hookers, or pretty much anyone you desire. In the early 2000s, everyday people were driving through crowds without a care—killing and injuring many. A kid in Austin, Texas drove through a crowd killing and injuring, he had no idea what had happened—like he awoke from a dream.

Call of Duty is a game where you shoot people with rapid fire and team up with people to perform these actions. People would have to wake up in the middle of the night to play this game. Then we have an epidemic of mass shootings.

When Fortnite went dark from the internet, people of all ages were angered and violent. This sound benevolent? This sounds like psychological conditioning for people to become what they are not—brainless killers. These video games are Trojan horses.

If you are skinny, then you're not fat enough. If your hair is straight, then you need to curl it. If you're too dark, then you need to be lighter. If you're corny, then you ain't dark enough. If you're too masculine, then you need to be more feminine. <u>If you are in the real world, then you need to be in an alternate reality—Meta.</u> Whatever you are, you need to be the opposite. Just to keep the children of all ages consciousness' occupied.

The main function of the Communist party is to control or destroy what they cannot control. All said games are mind control devices to destroy freedom and liberty from within. Communism is a cult of Xi Jinping and bans university studies on human rights.

The company Tidal Star indoctrinates its citizens in the ways and ideas of Xi Jinping president of the People's Republic of China and the Communist party who imitates Mao Zedong the founder of the People's Republic of China and the former ruler of the Communist party. Tidal Star is a company with daily lessons, compensated by the Communist Party. Just think of how it would be if a corporation in America forcibly train you to think like Trump or Obama. And you better get it right or find yourself in a re-education camp.

Mao Zedong was responsible for the 1958 to 1962 "Great Leap Forward" policy that led to more than 45 million humans mass murdered—topping Adolf Hitler and Joseph Stalin combined.

Why don't you know this? Because they weren't white enough for you too give a shit. Their eyes and language are too different from yours.

Only when you can see yourself in those mass graves do you care. And Israel and homeless people alike, play on your sympathy and emotions—they will not let you forget or think about anything else for that matter. They are the only ones who have suffered, right?

Lawyers who represent protesters are disbarred. Journalists who write critically about the Communist Party are thrown in prison. Xi Jinping, president of the People's Republic of China and the Communist party, has centralized authority by cancelling term limits of the presidency, reinforcing the idea of the Communist party leads everything. Xi Jinping is now king or Emperor to be more precise.

They have convinced the communist people to hang on and Xi Jinping will save us all, just keep hanging on. Or maybe your boss will tell you, just hang in there. We're gonna get you a raise, a new company truck, and lunch perdiem. Just keep hanging on, Trump will save us—reminds me of the church of Qanon. There is no Superman coming to save you —you must save yourself by understanding.

The pandemic started directly after the 5G startup, and people were being beaten by police for not wearing masks and dragged from their home to be placed in forced quarantine. Hong Kong's diminishment of the communist authority was the fuse on the powder keg that ignited everything we are experiencing.

In the early 2000s the Pentagon had a press release about the new non-lethal weapon the pentagon had developed by a military contractor. It used sound waves and radio frequency to disable their enemy—an energy weapon.

The Pentagon spokesman said if a foreign military lands on shore, they have the ability to disrupt their motor function, produce extreme fatigue, and give them flu like symptoms.

The reporter volunteered to experience this energy weapon. I did not see the reaction because I was listening to the broadcast on the radio—however the reporter couldn't stand the small amount of energy that felt like his lungs were being compressed.

The story was about this energy weapon that is also used in the telecommunications sector that was already being used. The question was, "<u>Is it safe in the hands of corporations or to use in the public sector?</u>"

The Telecommunications corporations said they would never do anything to hurt anybody, and so the Congress and Senate allowed it.

The answer is unequivocally NO. The telecommunications companies are draining us of our energy and lives.

Fast forward to the Hong Kong protest where protesters drove the Communist army back to Beijing with umbrellas and bow and arrows.

The first place to fully start up their 5G network was Beijing. Citizens in Beijing were ready to attack the communists because freedom and democracy spreads like a virus. There were reports of people losing motor function, extreme fatigue, and flu like symptoms.

We must stop this non-lethal weapon and draw back the energy being produced. Everyone is completely exhausted, can't sleep, and they have no idea why. The poor air quality and this non-lethal energy device that is used to speed up our communications is detrimental for convenience and hidden warfare. The telecommunications companies are guilty of aggressing on everyones' person and property.

Hong Kong needed guns to kill their oppressors. The only reason Americans didn't suffer the same Draconian tactics is because of a bunch of Right Wing nut jobs.

A corporate psychopath will not aggress even if there's a sliver of a chance they will be killed or lose access to their utilities.

Utilities - (Noun): something useful or designed for use.

Right Wing nut jobs are beneficial, but must be united under the Constitution and the Bill of Rights.

Rights are given to all humans at birth and are unalienable. They can not be taken or given away, it's just so. Black, white, yellow, red, brown, queer, straight, Christian, Muslim, etc. All free to live as they choose without aggressing on anyone else's person or property.

To enter a supermarket residents must show a green code on their phone through popular apps to purchase food. Someone with a low social score isn't permitted—more technological repression.

They tried something similar in America and were unsuccessful due to the 1964 Civil Rights Act designed to insure public service for all Americans. If you open your doors to the public the Civil Rights Act applies—it is <u>not</u> private property, it's public accommodations or black people would still be eating out back.

If you have trouble with entrance or you're being harassed for not wearing a mask—go to <u>constitutionallawgroup.us</u> and click on the Resources page—follow the instructions.

Hong Kong and Beijing haven't any weaponry to counteract this treachery—and the testicular fortitude has been chemically and psychologically reduced.

There are massive re-education camps for activists, lawyers, citizen journalists, and Uyghurs—a

Turkish ethnic group in central Asia—who are beaten, killed, and enslaved to produce products Americans love. More than ten hours of class with the same curriculum everyday—designed to change their minds, faiths, and beliefs—forcing them to renounce their religion.

International researchers found the Uyghurs and anyone else in the work camp were forced to produce Nike, Fila, Puma, Anta Sports Equipment, General Motors, Sony, Lenovo, Hewlett Packard, Dell Computers, Huawei, Apple, Acer laptop, Vivo Global, HTC Electronics, ASUS Global, Samsung, General Electric, Guangdong OPPO Mobile Telecommunications Corporation, Microsoft, and Meizu Technology Co., Ltd. products.

Basically you are trading someones consciousness for your convenience and entertainment. Kinda like when college students rape as an initiation - to steal someone's energy to up lift your own—Vampiric.

In this reality it's against universal law to take from one to give to another. If one can't survive on one's own, one disappears. If it has to steal consciousness to survive it's in violation of universal law.

Is there a finite amount of consciousness—only allotted to individuals? Do you have to steal to become more than? I think not. There is an infinite source, but it can only be accessed through light. Perversion makes it shy away.

The Smart City can identify Uyghurs, Muslims, or anyone who even speaks in private against the Communist Party with ease.

Capitalism can not exist with competition from slave labor, nor should humans be imprisoned because of religious beliefs that are not in alignment with the Communist Party. If black people can stand for slave labor—then fuck you. We must destroy these companies for providing this vile pain.

Huawei Chief Digital Transformation Officer Edwin Diender says it's a new way of policing the world—"Smart City." Diender claims he's human as well and is concerned the technology is used for espionage and totalitarian control. He's weighing his soul with the obscene amount of money and control he possesses.

Communist National Intelligence Law states "Any organization or citizen shall support, assist, and cooperate with state intelligence work according to law." China has no independent judiciary system for companies to appeal. All Chinese companies are communist wether they believe or not. It's a kingdom posing as a government.

Huawei has infiltrated most all countries and is building the telecommunications infrastructure—mysterious hidden cell towers. All designed for a back stage pass to the world and to control and exploit developing countries.

We in America live in the same police surveillance state. They're just not acknowledging it. The

Communists don't care what their subjects disapprove of, because they haven't any weaponry.

The United States uses it regularly and has crafty ways to gather other so-called evidence to help what they have already seen or believe they have seen. The technology America uses is flawed because they mostly get the story wrong, however they can gaslight you to fill in the blanks.

Beijing is afraid of the younger generation's power and should be. The best Beijing can do is taper down authoritarianism before their bloody destruction occurs. Hong Kong is in a police state now, soon it will be a better democracy than America.

China : Power and Prosperity was reported 9/30/2019 before the pandemic. Pandemic information was added after the fact.

Nick Schifrin with the support from the Pulitzer Center, and PBS News Hour empowered me with knowledge that helped me understand the chain of events that occurred. I knew the next move that was going to occur because of Nick Schifrin, the Pulitzer Center, and PBS News Hour—and I'm thankful.

Most of the information you read in this segment came from PBS News Hour China: Power and Prosperity, some from the Washington Post, the rest is my own experience.

Most other reporting from PBS is in alignment with the Communists in disguise as inclusiveness.

Now that I have explained to you how Communist China has imprisoned the largest population on the planet, I'll explain to you more about energetic attacks in Texas.

I believe HEB, Home Depot, Lowes, and other stores that are a part of the FEMA Emergency system have energy devices to deter non-compliant patrons by giving humans anxiety, confusion, stomach trouble, and are completely drained of energy causing people to pass out.

At the beginning of the mask mandate given by Texas Governor Greg Abbott, most people didn't wear one. And if people didn't wear a mask, HEB employees would hover around people and make human barricades while using their walkie—talkies in hopes the frequencies would effect people negatively.

One particular day a black lady dressed in a pink skirt suit—like she was going to church—passed me and I felt sick and anxiety with sweaty hands and feet. I found it very strange.

A few days later I drove to the Saltwater barrier in the North end of Beaumont, Texas and felt a sense of anxiety with sweaty hands and feet as I came to the railroad tracks at Helbig Rd and East Lucas Dr.— much like the Guadalupe River State Park.

The north end of Beaumont is predominately black people. The boat ramp at the Saltwater barrier had the same energy—all of the north end had the energy. Heading south on Pine Street I came to Interstate 10 and the energy subsided.

Later you hear on the news that minorities are mostly dying from the Corona/Covid virus. To be more clear I feel like the 5G infrastructure targeted an area with high crime and predominantly black people—to broadcast pain to the innocent and guilty.

Canyon Lake, Texas is a part of the San Antonio Metropolitan Statistical Area. It has the same energy as the Guadalupe River State Park and the Saltwater barrier in the North end of Beaumont, Texas. They are in areas with access to water and boat ramps.

Sunday, May 31, 2020, at 9:00 pm, a powerful frequency giving off a cigarette like smell made a powerful ringing noise in my ear. No one was outside smoking nor was anyone smoking inside my house. My TriField meter reading was peaked out meaning the highest frequency it could read. The frequency came from my wifi router.

Sunday, June 14, 2020, I was vacationing on Bolivar Peninsula near Galveston, Texas when—at midnight—I felt waves of peaked out cell frequencies coming from south to north. Wave after wave—my ears rang all night. The next morning my neck was locked up, I had TMJ, my lungs were inflamed, and not much sleep was had.

It comes more so when you sleep to be unnoticeable. It's more powerful when you're alone because you're the only one adsorbing the frequency, more people makes less concentration.

Is Mexico attacking us? Are the Russians that have many military bases in South America attacking us? Is there a boat giving off this painful frequency? Is it because the usage of cell phones was high all day and when everybody went to sleep the powerful unregulated cell frequencies just keep coming? Is it to keep people away from the coast? If it's coming from satellites we may be fucked.

People wake up exhausted—everyone in this wave of damaging cell frequencies had their person and property aggressed upon. The damages are at tremendous levels.

The most powerful frequencies are around areas with access to water. The telecommunications companies, on behalf of an unknown entity, are subconsciously driving humans away from access to water. If we are driven away from the coast, we are lost.

Just like the energy weapon that was used on American diplomats in Cuba, everyone felt it and received horrible symptoms.

In late 2016, US diplomats in Cuba began reporting weird and alarming events in their homes and hotel rooms. They spoke of irritating and piercing noises—squeals that weren't dampened when they placed their hands over their ears. Some described feeling pressure and vibrations as well.

Symptoms range from dizziness, nausea, headaches, balance problems, ringing in their ears, visual disturbances, nosebleeds, difficulty concentrating and recalling words, hearing loss, and speech problems.

More than 50 US diplomats and their families and more than 40 in Havana and at least a dozen more at the US Consulate in Guangzhou, China experienced this affliction. All have injuries to their brain. All in places where Communism exists—Havana and South America has Russian Communist, and you know about China. The government is now referring to it as the "Havana Syndrome."

I personally know people who also went to Mauriceville Elementary and suffered brain damage and memory loss in 2017—memory loss to this day. There are people in Beaumont, Texas who hear the squealing noise and now can also feel the boundaries and limitations of said noise.

The US State Department says it could be Radio Frequency Exposure. During the Cold War, Washington feared that Moscow was seeking to turn microwave radiation into covert weapons of mind control.

Now the US military has partnered with more than a dozen companies for "large scale experimentation" with 5G technology with efforts to enhance the "lethality" of certain systems and it will be a $600 million dollar project.

The military is calling it a "foundation enabler for all US defense modernization," and "dual application" says the Pentagon, which will give millions to private contractors to conduct tests at five US military instillations. Check the Department of Defense site. We must destroy these contractors.

In a speech before the United Nations general Assembly on September 24, 2019, UK Prime Minister Boris Johnson stated that digital authoritarianism is not the stuff of dystopian fantasy, but of an emerging reality.

A.I. and Smart Cities are a dark thundercloud lowering ever more oppressively over the human race, which the human race has no control over and there will be nowhere to hide.

Johnson asked if algorithms could be trusted with our lives and hopes and whether machines should be allowed to doom us to a cold and heartless future in an Orwellian world of censorship, repression, and control.

Donald Trump's Tweet, "I want 5G, and even 6G, technology in the United States as soon as possible. It is far more powerful, faster, and smarter than the current standard. American companies must step up their efforts, or get left behind. There is no reason that we should be behind on…"

Even Elon Musk said he had a meeting with all the governors in The United States about Artificial Intelligence and warned them that A.I. could not be trusted and people will die from this intrusion, and all ignored the warning.

I say we must kill this technology and its developers and operators immediately.

I have explained to you what I experienced as a child at Mauriceville Elementary and events leading up to now. I have the same symptoms these diplomats suffer—derived from the experiment in elementary school, and over exposure at state facilities. Believe what you see so we can all direct our energy to destroy the aggressors.

We know who the culprits are, and there indeed will be hell to pay. For every action there will be an equal and opposite reaction. We must destroy any telecommunications companies that are involved with this harmful technology.

It is more difficult to counter than a physical intrusion. The Military-Industrial Complex continues to develop and improve upon A.I. power over Americans. We must choke these bastards until we find the source.

<u>FIND THE SOURCE, SHUT IT DOWN, AND DESTROY IT!</u>

This technology is too powerful to be in the hands of anyone because absolute power corrupts absolutely.

XXII

Through all the troubles we have, everyone has forgotten the ability of the Russians to control and intrude into our very lives with ease through our communications network, our entire electrical infrastructure system, and every electronic we posses—without ever realizing.

Robert Phillip Hanssen who worked for the FBI from 1979 until his arrest was one of the many American spies who helped the Soviet Union.

Hanssen was a 25-year veteran and was arrested in 2001 for fifteen counts of espionage and plead guilty to all without the possibility of parole. Hanssen was receiving substantial amounts of money in return for classified information.

Dr. David Charney, the psychiatrist assigned to assess Hanssen, claimed that Hanssen could communicate with the KGB through his television set built into his cell wall, with the help of the infrared pulse coming from his remote control using Morse Code.

If this is true then the Russians have been technologically superior to America for quite a while.

Six Russian GRU officers were charged in connection with worldwide deployment of destructive malware and other disruptive actions in cyberspace. The defendants' of the malware attacks caused nearly one billion dollars in losses to three victims alone and sought to disrupt the 2017 French elections and the 2018 Winter Olympic Games according to the Department of Justice.

In October of 2020, a Pittsburgh federal grand jury indicted six computer hackers whom were all residents and nationals of the Russian Federation and officers in Unit 74455 of the Russian Main Intelligence Directorate (GRU), a military intelligence agency of the General Staff of the Armed Forces.

These GRU hackers and their co-conspirators engaged in computer intrusion with attacks intended to support the Russian government in efforts to undermine, retaliate, and destabilize the Ukrainian, Georgian, and French elections.

Come to think of it, we had problems with our elections, and have for some time.

Hackers disrupted efforts to hold Russia accountable for its use of weapons-grade nerve agents, and retaliated for banning Russians from participating in the 2018 Pyeong Chang Winter Olympic Games because of doping efforts.

Their computer attacks used some of the most destructive malware to date, including KillDisk and Industroyer, which caused blackouts in Ukraine.

Try going without power in the Ukraine, the blackout in Texas is chicken shit compared to the cold this region experiences.

In December 2015-2016, the destructive malware attacked the Ukrainian electrical power grid, the Ministry of Finance, and State Treasury Service using BlackEnergy, Industroyer, and KillDisk.

On April and May of 2017, there were hack and leak efforts targeting French President Macron.

In June of 2017 the destructive malware infected computers worldwide using malware known as NotPetya, including hospitals and other medical facilities in the Heritage Valley Health System in Pennsylvania, a FedEx Subsidiary, and a pharmaceutical manufacturer.

In 2018 Russian hackers targeted a major Georgian media group, then in 2019 Russian hackers compromised Parliament networks and a wide range of websites.

Cybersecurity researchers tracked the Conspirators and their malicious activities using the labels "Sandworm Team," "Telebots," "Voodoo Bear," and "Iron Viking."

No one is protecting us from cyber attacks and no one is the lead on cyber attacks. We must unify to protect ourselves from this destruction and disruption, or we can just leave the cyber world all together.

Through Facebook, the Russians were successful at convincing white Americans that Russians are the good guys. Qanon also helped this idea of Putin helping Trump destroy the Shadow government.

People are imprisoned for criticizing Putin; just because they're white doesn't mean they're on our side. Everyone wants America to fall because of the demons they have suffered from our government.

It appears that the Democrats, Russians, the media, and Facebook confused Americans on whether the Sandyhook shooting and many more shootings were staged with crisis actors.

News papers across the country used Aurora, Colorado theater shooting suspects defense attorneys Iris Eytan's picture, or look a likes in other shootings and crisis—showing her screaming and crying in the street holding bloody shooting victims. After the confusion, most Americans didn't believe any shooting or other events were real. The media is who put doubt in the hearts of Americans.

I have seen Mr. Peel's Deep Fake videos making Presidents of the United States use words and facial gestures controlled by Mr. Peel. I saw the Deep Fake videos that showed the parents of the school shooting victims acting like they were preparing for their lines to perform in front of the camera. Laughing and asking what they should say, and then crying while on camera.

This didn't clear anything up—it caused more doubt. The evidence of Deep Fake is clear, but when and how it is used made everything fake.

I don't blame people for not believing because of the false evidence, and the heavy attempt to take guns away from Americans failed. Demons twisted the very fabric of our existence.

We didn't know anything about Deep Fake before Mr. Peel's explanation of it. The Russians are obviously technologically more advanced than we. Deep Fake and Sandy hook is one of the ways America became weak as a nation, and the mere sight of a gun causes paralysis.

They damn near had Texans attack the military during the Jade Helm event. We were ready to cut the military a new ass and the governor had Texas Rangers question and examine the situation.

We must remember that most of our military volunteered to protect Americans, our way of life, and the greater good. Their wanting to be Rambo, or Seal Team whatever could distort their judgment, but they should always be reminded of what they are really designed to do. Much like the police, they will protect our unalienable rights of the Constitution of The United States or suffer the consequence.

This all came from Facebook and the ability of Facebook to accelerate people's emotions with a technologically advanced shock collar, designed to socially engineer you. It was designed to ensure more gun control—it failed.

The 5G can accelerate and agitate people at demonstrations to become violent, and in this instance was designed to cause violence to ensure gun control.

Without weaponry the Communist corporate psychopath will have full control over all humans.

The real question is, were these events orchestrated by the Russians, Democratic Party, the media, or Facebook whom are in direct alignment with the Communist Party in hopes of taking guns way from Americans? The Russians, Chinese (which are communist-voluntary or in-voluntary), and the Democratic Party appear to all be hell bent on destroying freedom in disguise as safety.

We have so many enemies that it's hard to know. I blame the CIA, FBI, and alike entities for their demon actions against human beings. They have caused so much pain and suffering that we now all suffer.

The hacker group called the DarkSide used ransomware to disrupt, shut down, and steal money from the Colonial pipeline. For doing so, the FBI seized all their crypto Bitcoins. Calling it crypto is false advertisement.

Don't you find it amazing that law enforcement can seize your intangible digital fiat money thats isn't backed by anything? It doesn't matter if you have a fifty-thousand numbered digital code, if you live in this alternate reality it's controlled to the fullest extent. This is what they want to start using to vote, and most people still believe that it's the safest way to live their lives.

In no way should something so valuable ever be subject to this Trojan horse.

It has been said that the FBI and the media is lying about the seizure, and it's impossible to seize crypto currency. Who the hell can tell anymore? All said entities' creditability is diminished. Can Colonial Pipeline remember what a valve wrench is, and can they pull their heads out of their asses? All tough questions.

CNA, a Chicago based financial corporation among the largest insurance companies in the U.S., paid a $40 million ransom.

No one is protecting us, and by paying they are setting a precedent of success. For all we know they all hacked themselves.

Conditioning the Brain

XXIII

I've been working on the railroad—all the live long day. For how long? Till Gabriel blows his horn—to the end.

This was the first song I learned when I came into this world. It's a folk song my great aunts taught me to be playful with children—after all they grew up in the 1930s.

Work your ass off until the end, and do it for the team. Growing up I understood if you work hard for the company you'll be rewarded. I'm not sure that has ever been true. Yet a feeling of worthlessness comes over people when they are unemployed—deep conditioning.

I enjoy working hard. It brings me satisfaction. As I get older, the satisfaction of hard work only comes when it's beneficial longterm.

We are all drawers of water and cutters of wood. We must physically work to stay healthy in mind, body, and soul. There's no way around it in this reality.

Having an occupation looks as if we are being occupied to keep us from turning on our masters. Our masters want to keep blinders on our consciousness to keep us from noticing the cage.

Your enslavement started with the threat of pain. Humans' fear of death made them controllable. When the mind is controlled, the body takes to the shackle. Human beings are the greatest natural resource on the earth. Like most natural resources on Earth, people of power take control.

Human Farmers (corporations that own humans as products) own human livestock and demand productivity from their possessions. An invasion of the mind is how they involuntarily control the livestock with techniques like industrial management and psychological assessment.

Deception of the livestock for personal gain is a crime against humanity and the natural laws of the universe. Human Farmers have aggressed on everyone's person, property, and piece of mind.

The first control is to indoctrinate humans through education. If you control the thoughts and souls of the livestock, the livestock usually won't understand they own themselves at birth, and have the absolute right and ownership of their person and property. Your person and property are one and the same, yet at the same time they are two different entities.

Your person is your mind, body, and soul. Your property is also and simultaneously your mind, body, soul, and previously unowned natural resources that you find, transformed by your own labor, and then given to or exchanged with others <u>voluntarily.</u>

If you don't know you own yourself from birth then you are an unclaimed resource. The Constitution is written documentation and identification of what you already own from birth. The founders of the United States didn't give you shit. They only wrote down what everyone already owned from birth.

The Educational System is the most insidious way to control impressionable children to become compliant employees. It has been the central component to Communist and Fascist tyrannies through history. The book "The Deliberate Dumbing Down of America" exposes the role of Globalist foundations.

Another way to occupy human livestock is to turn the livestock against each other. You can be a Democrat or a Republican, a conservative or a liberal, you can be with them or us—it's called "Controlled Opposition." Religion is the backdrop of nearly every war throughout history.

Your masters want you to believe there are only two choices—winners and losers. This will keep anyone from gaining ground against the oligarchy.

Oligarch - Greek, meaning few to rule. A very rich business leader with a great deal of political influence.

The central theme is the same throughout : Divide and Conquer. The Globalists short circuit the natural tendencies of people to cooperate for survival and teach them to form teams hell bent on domination and winning.

Sports has always had a role as a key distraction that corrals tribal tendencies into non-important events. Modern sports has reached a ridiculous level where protests breakout over celebrities leaving their city, but essential human liberties are giggled away as inconsequential.

The roman poet Juvenal writes "Give them bread and circuses and they will never revolt." Sports is the modern day Coliseum.

Being on your own and having your own thoughts is too risky for livestock. The livestock owners must have dependent livestock to remain successful—dependent in thought and sustenance.

We don't breed well under captivity. That cuts down on the livestock head. You must believe you're free to drive the livestock numbers higher.

Humans are the top resource and are treated as if they are useless. Nothing can work or continue without humans. The more supply of livestock, the less valuable they become. The more subservient livestock there are, the easier it is to fill the gap if some of the livestock turn on their masters. There is always someone ready to fill the gap, especially if the area is over populated and poor.

If someone is desperate because of a lack of resources they can claim—food, water, shelter—they will become more willing to do what ever is necessary to survive because of the fear of death.

A bigger population needs a bigger grazing field or bigger cage.

It's hard to control livestock by force. It makes them under productive. They must believe they are free—free range humans. Your masters will let you pick your own occupation—you can work the field or if you're worthy you can work in the house. The human farmers raise your productivity so you can pay your master—it's called a Tax Farm.

I find it funny that someone would surrender a portion of their property to furnish the means to protect the rest.

Binary hegemonic intervention occurs when the invader forces the subject to make an exchange or a unilateral "gift" of some goods or services to the invader—with the threat of imprisonment or violence.

Even though taxes are voluntary, someone will come to collect this coerced gift with the point of a gun or the theft of your property for not supplying said gift.

Tax Inspectors Without Borders is an organization to help tax collectors around the world figure out how to collect taxes where there is no money; to hold "Stakeholders Workshops" and be the frontline in the battle against tax avoidance. To say "without boarders" gives it a humanitarian feel to it, like Doctors without Borders. They are the front line in this coerced gift because they are at war with us.

They must put some of the livestock on the payroll—such as police and bureaucrats. From desperation, the payroll livestock will attack anyone who questions the farmers that farm humans.

Questioning the farmers that farm humans could interfere with how much feed the payroll livestock will receive. And they will kill and pollute if necessary—because of the fear of death.

Intellectual and artistic livestock are forever dependent on the farmer and will indeed harm other livestock if a threat of feed is reduced. You can reduce the feed and tell the intellectual and artistic livestock the reason they didn't receive feed is because of agitators or the unvaccinated. This will direct their anger at anyone who questions the Human Farmer.

The most effective ploy that I have witnessed is a "continual threat." Terrorists, drugs, killer bees or wasps, a great number of flus from birds, cows, swine, mosquitos, purple people eaters…

Next will be an alien invasion or the holographic coming of Jesus. What ever they can do to keep you afraid and occupy your consciousness is what's next. The farmer will provide a continual external threat so the livestock will cling to protection because of the fear of death.

"It's the leaders of a country who determine the policy, whether it's a democracy or communist dictatorship. The people can always be brought to the bidding of the leaders. All you have to do is tell them they are being attacked."—Herman Goering at the Nuremberg trials.

Once a monopoly of protection is established, coercive autistic intervention follows. The individuals who are coerced into saying or not saying something or into making or not making an exchange with the intervener or with someone else is having his actions changed by a threat of violence.

Autistic intervention usually comes in the form of homicide, assault, and compulsory enforcement or prohibition of any salute, speech, movement, accessibility, or religious observation.

The intervener is the State which issues the edict to all individuals in the society.

Edict - an official order, mandate, or proclamation issued by a person in authority.

US Supreme Court Marbury vs. Madison 1803: "Any law repugnant to the Constitution is null and void." It has no force of law, and can be ignored.

Mack Printz vs. United States 1997: Justice Scalia wrote, "The Constitution protects us from our own best intentions." Any code the government claims to pass "for our safety" cannot be upheld or enforced since it is outside the rule of law and governmental authority.

Therefore, inform the government official that you will sue them personally; their government status does not protect them if they fine, close your business, or arrest you.

Under TITLE 18, USC SECTION 242, DEPRIVATION OF RIGHTS UNDER COLOR OF LAW,

TITLE 42 USC 1983 - CIVIL RIGHTS ACTION FOR DEPRIVATION OF RIGHTS,

TITLE 42 USC CODE 2000a - PROHIBITION AGAINST DISCRIMINATION OR SEGREGATION IN PLACE OF PUBLIC ACCOMMODATIONS,

TITLE 42 USC CODE 1985 - CONSPIRACY TO INTERFERE WITH CIVIL RIGHTS,

CIVIL RIGHTS ACT OF 1964.

It is a crime for a person acting under color of any law to knowingly and willfully deprive a person of a right or privilege protected by the Constitution of the United States—punishment for said crimes will be applied.

No shoes, No shirt, No service is a Decency Law and has nothing to do with discrimination. Opening your doors to the public gives them an irrevocable public license to enter any place. When you open your door to the public you are a Public Accommodation.

The police officer that doesn't inform or if they remove you from said public accommodation will be held personally and professionally liable in Federal Court. Collect their badge number and name, it's best if you have them on camera. It is treason to not uphold the constitutional oath they must all take. It is their job to uphold the Constitution of the United States. Treason is going to be a painful lesson for them and many more. Treason is punishable by death.

Servitude works best on a voluntary basis—if it is not voluntary then it's coercion and slavery.

It should be easy for you to see the modern police state and the growing snitch culture wrapped in a pseudo—patriotic War on Terror, a virus, or an alien invasion we must all come together to over come.

The consolidation of media has enabled corporate structures to merge with government which utilizes the concept of communist propaganda placement. Television, print, movies, cable news, and local news can now work seamlessly to integrate an overall message which seems to have a ring of truth because it comes from so many sources simultaneously.

Leopold Tyrmand's 1976 essay called "The Media Shangra-la, Where everyone is Free but some are Freer;" printed in the American Scholar quarterly magazine by The Phi Beta Kappa Society says the polished totalitarian gimmick of communist and Nazism (National Socialist Zionist) is efficiently applied.

"A frequently repeated lie isn't truth, but it survives and no one knows the difference. The American press applies this trick to a more sophisticated use than Communists or NAZIs ever did. Instead of repeating the lie the press has endless versions of it."

When you become attentive and can identify the main message, you will be able to see the imprint everywhere—not counting sub-liminal messages.

Television programming is engineered to literally lull you to sleep, making it a psycho—social weapon.

Flicker rate tests show alpha brain waves are altered, producing a type of hypnosis—which doesn't portend well for the latest revelation that light can transmit coded internet data by flickering faster than the eye can see. Satan masquerades as a messenger of light.

Flicker rates from computers, video games, and social networks provides a basic structure which overloads the brain with information. The rapid pace of modern communication induces a state of ADHD.

This coordinated script has been in place for a long time. It's goal is to turn humans into non-thinking automatons.

"The best way to turn willing participants of a free nation of independent citizens into slaves is with slow imperceptibility—without violence—to corrupt them, demoralize them, deprave them, until they have acquired the characteristics of slaves."

Just as Circe turned Odysseus's men into grunting swine, the media waves their weapon of mass distraction and turn humans into automatons, living machines.

You can read about this technique in the prestigious Jewish Quarterly, 2004. This essay spelled out in detail how they manipulate the billion dollar porn industry. Most of the people who read this couldn't believe their eyes. I told them all "it's in front of your face—in black and white."

It's the only way the small "elite" can bend the large population. A perilous phase where mind control has taken on physical and scientific dimensions that threaten to become a permanent state if we become aware not of the disposal of the technocratic dictatorship unfolding on a world scale.

Through propaganda, MSM and CNN has turned white liberal women from being anti-big corporation to worshipping big pharmaceutical. Also, gays have gay pride floats sponsored by Goldman Sachs. The weak-minded are always manipulated for social discourse.

Consumption - the using up of resources, a wasting disease.

Edward Bernays was a pioneer in the field of public relations and propaganda—also the inventor of the consumerist culture. It was designed to target people's self-image in order to turn wants into needs. Bernays noted in his 1928 book "Propaganda" that propaganda is the executive arm of the invisible government.

The Marine Lt. Col. Oliver North tried to excelled this shadow government in 1984 by an attempt to impose martial law. North secretly planned to suspend the Constitution and allow at the time a little known Federal Emergency Management Agency (FEMA) to control operations that would have been accountable to no one and appoint a military commander to run state and local government.

North was going to have President Ronald Regan—who was in decline from dementia—impose this action through executive order. Sound like something we are experiencing now?

You can read about this in the Chicago Tribune—July 5, 1987.

The war hero Senator Daniel Inouye from Hawaii says—yes—there is a shadow government.

Senator Daniel Inouye fought in World War II with the 442nd Regimental Combat Team for which he received the Medal of Honor.

Senator Daniel Inouye went on to state this shadow government has its own Air Force, Navy, its own fundraising mechanism, and pursues its own ideas of national interest, free from checks and balances and free from law itself.

Oliver North claimed he wanted to perform operations forbidden to the CIA and fund this shadow government by generating money through US arms sales.

This is an unseemly disregard of our laws and Constitution of well intended patriotic zealots who adopted the Marxist views of "the ends justify the means." Is it because of the inadequacy of our laws and Constitution? Most likely it is the inadequacy of the national leadership.

North and Admiral Poindexter both stated "this is a dangerous world." This is an excuse for autocracy and their psychological conditioning from an early age to become communist—and did.

Autocracy - a system of government by one person with absolute power.

Our form of government is what gives us strength. It must be safeguarded particularly when times are dangerous and the temptation to aggregate powers are greatest.

Aggregate - form a separate unit.

Vigilance abroad doesn't require us to abandon our ideas or rule of law. Without our principals and ideas there is little that is special to defend.

In 1787 when the National convention finished its business—a by stander asked Dr. Benjamin Franklin, "Well Dr. Franklin, what do we have? A Republic or a monarch?" Franklin replied, "A republic if you can defend it."

The CIA must be destroyed and dismantled along with FEMA. Oliver North, John Poindexter, and all NSA staffers involved must have their liberties arrested immediately or their death by brutal, blunt force.

They are in direct alignment with the Communist Party that will stop at nothing to have Americans destroy themselves from within. It is no surprise to me that the media turned Oliver North into a television hero. Propaganda was and is powerful dark magic.

<u>The most important hurdle for the Communist Party is to have Americans abandon the Constitution.</u>

Harvard 1962, the CIA commissioned Henry Murray to create a "Manchurian Candidate" for MKULTRA Project by conducting humiliation experiments on unsuspecting freshmen. Among the participants were Ted Kaczynski (the alleged Unabomber), Timothy Leary, and Ted Bundy.

Henry Murray was a pioneer of personality tests for industrial management and psychological assessment. When I say industrial management, I mean conditioning your mind for compliance to perform actions contrary to health and wellbeing.

The class you would take to understand some of the aspects of this psychological control is called <u>Behavioral</u>—the new technique to control employees at a corporate level under the philosophy of the Communist party.

Henry Murray was an OSS Chief psychologist (Office of Strategic Services) who monitored military experiments on brainwashing and sodium amytal interrogation.

In 1988 Henry Murray died and was sent back to hell. This demon was a high priest in the new religion called Science. In this religion it is good to destroy other living beings in the name of Science. They need blood offerings to gain great power.

The earth-men called upon the Kingdom of Shadows to gain great power, and the Children of Shadows were called up from deep below us. They were formless of another vibration, existing unseen by the children of earth-men. Only through blood could they have formed beings. Only through man could they live in the world. When blood was offered, they came to dwell among men.

Only by sight were they men. They were serpent-headed and when the glamour wasn't lifted, they appeared to be men among men. And they crept into councils, slaying chiefs of the kingdoms. Only by magic could they be discovered, only by sound could their face be seen. Their mission is to destroy man and rule in his place.

They taught man the WORD that only man can pronounce, the serpents still liveth. These ancient words ring true still.

Like when they gave 600 unsuspecting impoverished sharecroppers from Macon County, Alabama syphilis and claimed they were treating them for the said disease. They were told it would only last 6 months when in fact it lasted 40 years.

The experiment was called the Tuskegee Syphilis Study. This was a willful destruction of one human by another. Some became insane and lived painful lives. Mostly it was to destroy the black community. After all, they were just a bunch of poor, dumb niggers—this is how demons operate.

Now there is a historical class-action lawsuit against the CIA's illegal use of mind control.

According to CBC (Canadian Broadcasting Corporation), over 40 Canadians were victims of MKULTRA at McGill University and Allan Memorial Institute in Montreal, Quebec which is a part of said University.

These experiments were supervised by Dr. Ewen Cameron. The main focus of this experiment was the "depatterning" and "psychic driving" to erase memories and reprogram them with new thoughts. If they don't like what you think they will just change it like a robot or automaton—thought police. The demon Cameron went back to hell in 1967.

The Canadian government paid millions to Allen Memorial Institute in Montreal with a cover organization called the Society for the Investigation of Human Ecology.

The book "Brave New World" foreshadows medical tyranny and has disorders for everyone—particularly those who question authority. Nerve drugs in the military have led to record numbers of suicides and more than 25% of U.S. children are on mind numbing medication.

Patients started the program with insignificant mental health issues—depression and anxiety. Dr. Ewen Cameron performed tests with drugs like LSD and PCP—medically induced sleep for extended periods, and over saw extreme forms of electroshock and sensory deprivation according to the CBC.

These experiments included a little girl of 7 years named Ellen Atkins. Her mother Emma Jane Crunican was in the sleep experiment in Montreal in the 1950s. Suddenly her whole family was being treated like guinea pigs for the sake of power over the human mind.

The new religion, Science, destroyed many people's lives. Science has been infiltrated and taken over by demons, and the universal truths are diluted to a point where no one knows what is true anymore.

Instead of claiming you only trust in science you should just say you don't have critical thought, and you default to believing whatever most other people think to prevent the feeling of inadequacy.

If you feel that you have experienced what is being described, start by contacting Survivors Allied Against Government Abuse. If it's a CIA cover to trap guinea pigs—fear not. The CIA and its employees' days are numbered.

We may need to provide psilocybin with guidance to relieve us from our conditioning and lift the veil—nature and the Lord provides.

As I described earlier, personality tests for Industrial Management and Psychological Assessment (Behavioral) has assisted in cases where the evidence of damages is empirical. The victims are induced with

Stockholm Syndrome and feel a sense of compassion for their captors.

The TPC plant exploded in Port Neches, Texas damaging many homes. It exploded because of extreme willful negligence. The owners of TPC didn't even care about the public school that was destroyed. By the grace of God, there were no students that day.

The many residents that received damages were convinced that TPC could fix their houses and it would be reasonable to ask for 10% over the cost of damages. TPC convinced the victims that if they sued, the plant would close down and there would be jobs lost. Nothing could have ever been further from the truth.

No plant will shut down, they can't. The tangible assets are too valuable and they can't pick them up and go somewhere else. We have them over a barrel and they are fucked, and it's time to pour the coals to them.

Through personality tests for Industrial Management and Psychological Assessment (Behavioral), they knew the "white trash" would never stand against their masters.

Like a bunch of slaves they fell for it and didn't cause TPC anything meaningful, even though the evidence of damages against person and property is clear and obvious. They should have demanded $5 million a piece and sued individually, fuck a class action law suit.

But something tells me after reading this and ascending from Plato's Cave, that will change.

A corporate psychopath only fears two things, the loss of utilities or the fear of death. In this case the corporate psychopath didn't lose any utilities nor did they have the fear of death. So that means business will be conducted as usual, and nothing meaningful has been produced, yet.

In the corporate world you want corporate psychopaths to believe you're participating even if you're not. The fear of having your resources, utilities, or disapproval is too great. Your years of conditioning cause your Stockholm Syndrome to flair.

A way to help alter your conditioning when you identify a corporate psychopath is first think to yourself, "I may have to kill this asshole." This will greatly help alter your conditioning.

Corporations like HEB stopped the transportation of food to help efforts to allow immigrants to enter the United States and to induce fear for profit. HEB claimed there was no one to transport the produce and supplies. There are plenty of drivers, just not drivers who will accept dirt-cheap immigrant wages.

Corporations like HEB and Whole Foods enjoy exploiting immigrants and buying extremely cheap produce, all disguised as humanitarian efforts. Propaganda presents you with a lush green organic picture, but their greed is showing through.

One of the most important reasons we need not anymore immigrants is because they are compliant to any order given to them out of desperation. They don't come from people with liberties and freedoms, they lack the ability to comprehend freedom and liberties because of generational slavery.

They are diluting our freedoms with slave mentalities. They are people who have been institutionalized by communist/socialist government and have outsourced the rearing of their children to said State. They are people of compulsory education leeching from state funds and are also conditioned to comply.

They care not what freedom feels like, they only care about the economical benefits. Mexican immigrants believe they own American land anyway, and want to take it back without firing a shot. They want to take it by occupying it. We must obstruct immigrants from any country that doesn't already have at least real Freedom of Speech.

By the way, Canada doesn't have Freedom of Speech and is under communist authority. Dr. Dena Churchill was fined $100,000 for criticizing vaccines on social media.

Legislation must be implemented on the safety and accessibility of our food. In almost no circumstance should anyone be prohibited from access to food and water.

If there is a scare and said grocery stores don't have the ability to transport or exhibit food, then we must implement brave citizens to help with efforts and the cost will be adjusted for or against the grocer. If employees are afraid to work then they aren't being paid enough and should receive hazard pay, and/or brave citizens will be called to action.

Any elected officials that even suggest a "vaccine passport" will find their lives mysteriously going to shit and a great deal of pain will be inflicted. It's funny to me how most people on social media are ready to stand against the NAZI, and white people are in love with World War II because of the evil they want to stand against. While they go to the grocery store and the Gestapo asks "Papers please," and they freely submit. Your ignorance exceeds.

In no way should a fat lady with poor hygiene that works at a grocery store, coffee shop, or restaurant ever give orders to patrons, only take them. It is none of their business if someone wears a mask or not, everyone is free to do as they choose.

Wearing a mask is a way to signal to other like-minded people of their tribe and has nothing to do with safety from sickness. People will continue to wear masks for years to come in defiance of former President Trump and support for the Democrats. I like it because you can identify their level of consciousness, only if we could do the same with the church of Qanon.

It's obvious these fat ladies are not in control of their own lives, nor should they be in charge of anyone else. I guess like Black Lives Matter and LGBTQ+, they finally had their day in the sun, now these ladies had their 15 minutes of fame as well. I'll be glad when we become humans again, as opposed to separation of said color or creed.

Covid has psychologically divided us, while unifying us in a cyber surveillance dystopia. To unify us on a moral high horse with a social credit score clearly seen with a mask of who is participating. It's all fear based trauma, and is clearly seen when people grocery shop with clear buckets over their head, while wearing a mask.

Tech companies forced people to use their product by inducing fear of going to the grocery store and even produced physical symptoms. Pay us and stay away so you can use the grocery app and we'll know your wants and desires. That technique hit deep in the subconscious mind of most all in the world.

Hell, the only company that could stay in business during the so-called pandemic was Amazon.

The pandemic is speeding up automation and more that 85 million jobs are on the line. Bank tellers are being replaced with robotic tellers. This is job deconstruction and the return to serfdom.

Small town small businesses can't compete with tech communism, and it disrupts capitalism with an unlimited amount of money being paid. They price people out of their homes and communities, then the greed of taxes finishes them off. This is one of the many ways communism is disrupting true capitalism.

Small business were the only stores being burned down from riots, no Wal-mart burned down. These rioters are controlled by communist factions to destroy small town capitalism, so that we will be at the will of the corporate communist psychopath.

Corporate Capitalism is based in desire—if you want it, it costs more. If no one wants it, it ain't worth shit, a Bernays style of turning wants into needs. Socialism is based on control by force, if it ain't voluntary—it's by force.

Through propaganda, Corporate Capitalism has even made cancer profitable—just think of all the people they have poisoned back to health. A continuous threat with continuous awareness, walks, and even buckets of chicken wings. All the KFC chicken buckets, link sales, 5K walks can change not cancer. Clean air, food, and water along with the prohibition of electromagnetic fields will.

Just after 9/11, the media was testing the effectiveness of their psychological conditioning. On an episode of The Rosie O'Donnell Show, with guest star Natalie Portman, the show emphasized the death of actor Timothy Hutton, who has been on five more shows since 9/11 and is living in California as we speak.

Hutton played opposite to Portman in the Film "Beautiful Girls." Rosie O'Donnell emphasizes the false idea of Hutton's death and wanted Portman to confirm. Portman simply stated, "I don't want to do that," to me this says she didn't want to be a part of the lie. This is all prior to social media, which took the lead in psychological conditioning. The misinformation was accepted and the test was successful.

Prior to 9/11 all radio stations in Southeast Texas claimed country singer Don Williams died as well. About eight years later Don Williams performed in town and people were surprised to see him. I knew people who went to the concert just to see if it were true. They asked Williams about the said death and it caused him great pain, it was clear to see in his eyes as people spoke to him. Don Williams said he would never come back to Southeast Texas.

There were more, so many more. Was this a test of misinformation or a way to punish the said performers?

Reality shows pick contestants that can be manipulated for outbursts, it makes good TV. Reality shows have ruined many lives with the psychological manipulation to perform involuntary actions viewed as low character qualities and later the contestant has trouble finding a job and are shunned from society. The fact that it was involuntary is troubling.

How many ways have our news sources turned us against truth, freedom, and liberty? That number is uncountable, but is coming to an end.

Sharyl Attkisson's book "Slanted: How the News Media Taught Us to Love Censorship and Hate Journalism," explains the misreporting of many issues like Black Lives Matter, Corona Virus, Joe Biden, Silicon Valley censorship, and much more.

Every brand name outlet you read or watch is full of shit, and designed to turn from freedom to communism. The media is now saying it because in this universe everyone has to tell you what they are doing to you. Listen to them and believe them.

Doesn't it feel better watching television now that Trump is out of office? That was mighty white of them to let us out of hell. Every media outlet induced fear, pain, and anxiety to watchers and followers, they were too scared to turn it off. The unknown death was around the corner, just look at the television, you'll see. They applied the fires of hell on you so you would do anything to change the circumstance, including elect a fuckin' rubber boot if they would just stop hurting you.

Project Veritas sued the New York Times in a defamation lawsuit and the New York Supreme Court did not grant motion to dismiss.

The New York Times, along with coordinated media outlets, claimed Project Veritas doctored their videos and the report on voter fraud was disinformation, even though the video was produced by the criminal himself. The video showed ballots, money, and elderly people whom he paid. There were many more people who came forward with videos as well and risked their lives to expose voter fraud.

The New York Times relied on Wikipedia's opinion of Project Veritas, much like Wikipedia's opinion of Stephan Molyneux, for reporting and cited it as a defense in their lawsuit.

Justice Charles Wood identified the reporters of The New York Times interjected their opinion, all the while The New York Times claiming they were *"libel-proof."* That is at an end.

Justice Woods said, *"polling does not decide the truth, nor speak to evidence,"* and *"defendants have not met their burden to prove that the reporting by Veritas in the video is deceptive."* Woods went on to say, *"defendants acted with actual malice, that is, with knowledge that the statements in the articles were false or made with reckless disregard of whether they were false or not."*

Bravo Justice Woods!

This is a major victory for truth against opinion and will be a deep cut for infringers and legal protections of infringers.

The Sullivan decision was a landmark US Supreme Court ruling that restricted the ability of American public officials to sue for defamation. It's like saying we didn't know we were wrong, so I'm protected. We must strive for the truth to cut down on the chatter and the dismantling of our society.

All coordinated media groups claimed Project Veritas is a disinformation source, and had major social media groups send out tens of thousand of notifications to its subscribers claiming so.

It is proven by the New York Supreme Court that The New York Times is producing opinion with malice. USA Today uses the New York Times as its fact checking source, and Facebook uses USA Today as its fact checking source. All said media groups did what they claimed Project Veritas was doing, it's called Psychological Projection.

Psychological Projection - a defense mechanism in which the ego defends itself against unconscious impulses or qualities by denying their existence in themselves and attributing them to others.

The fact checking phenomenon started around 2005 when everyone started using alternative news for information. Fact checking is a maneuver of retaliation from main stream media. They have a monopoly over information and want to keep it. There must be significant consequence to propaganda and news as entertainment.

Rupert Murdoch sensationalized the news as entertainment and was used for propaganda purposes—this will be corrected.

Murdoch owns News Corp which owns hundreds of local, national, and international publishing outlets around the world. These outlets include—(UK) The Sun and The Times, (Australia) The Daily Telegraph, Herald Sun, (US) The Wall Street Journal, New York Post. Murdoch also owns book publisher HarperCollins and television broadcasting channels Sky News Australia and Fox News (owned by Fox Corporation) and until 2018 Murdoch owned 21st Century Fox.

Rupert Murdoch is an Australian, and Australians are raised with communist characteristics.

We know this to be true because their weapons were taken from them. Murdoch is one of the reasons we don't know what is true anymore. We must punish him for this.

Project Veritas also caught Charlie Chester, the Technical Director of the news outlet CNN, on video stating that CNN boosted the Covid-19 numbers, and swayed reporting to help Black Lives Matter. One of the many ways to <u>not</u> report that black people were producing the Asian violence and more. CNN wanted it to look like white supremacists were aggressing on Asians' person and property.

Chester specifically went to work for CNN to sway the election and stated on video that CNN produced propaganda.

People say the darnedest things after a few drinks and the promise of some pussy; both do confuse the mind if the conditioning is in order.

CNN is owned by AT&T and CEO John Stankey says that, "CNN is the most respected brand in news today. We are proud of that." Stankey drank the kool-aid and now it's time to lay down quietly and pass away.

A CNN paid contributor and former Republican Senator, Rick Santorum, says, *"We birthed a nation from nothing. I mean, there was nothing here. I mean, yes we have Native Americans, but candidly there isn't much Native American culture in American culture."*

Well, Rick, that's because in 1869 hundreds of thousands of Native American children were removed from their families and involuntarily placed in boarding schools. They were forced to give up tradition, language, and their cultural identity with oppression and violence. The U.S. government implemented this assimilation to make them as much of a slave as everyone else. The children's homes existed up until recently.

CNN pays Santorum a six-figure salary to vomit anything that will come out of his pie hole to fill time. CNN knows America has forgotten about the true owners of this land and want to continue doing so.

CNN is selective of what they care about for ratings and are indeed psychopathic because they only care about themselves, or what they are conditioned to care about. Media outlets that lack independent thought are a danger to freedom.

Here's a funny little way to rebrand bad acronyms with good ones.

In 2014, the Bureau of Land Management (BLM) was under fire for fighting against ranchers in Nevada. It looked as if Senate Majority Leader Harry Reid and his son wanted land that was being used as free range grazing. The deal was the Communist Party would produce a solar field in competition with a coal plant nearby, and they wanted cattle off the land.

Later on Facebook, videos arose of BLM shooting one of the ranch workers in cold blood on a highway. The strange part is they had video inside the vehicle of the ranchers corresponding with aerial videos, as if to be produced to enrage Americans.

All this took place at the same time as Sandyhook and Jade Helm—a time when the attempt to enrage Americans to produce gun violence so they can produce more gun restriction.

A short while after, we were on the verge of a civil war—black against white. Then a solar eclipse came and just a few days later all of America was flooded, it extinguished our fire and violence. The next thing you know hillbillies in monster trucks were saving little old black ladies from the flood, and vice versa. Now BLM is rebranded as Black Lives Matter… weird.

The invasion and control of our consciousness runs deep. So deep that a man in Boston, Massachusetts on New Berry St. was almost knifed at a Starbucks for not wearing a mask during the pandemic scare. The aggressor at Starbucks thought he was protecting people by harming another. It's called the "Bystander at the switch."

In 1948, the United Nations formally adopted the Universal Declaration of Human Rights after the Second World War due to the human rights violations committed during this period.

It explicitly forbids governments from treating some humans more valuable than others or from sacrificing some for the benefit of others.

It also forbids governments from knowingly imposing harm on some individuals in order to serve an alleged "greater good," and forbids governments from imposing a hierarchy of rights on their citizens. Had it not been for this, NAZIs (National Socialist Zionist) would still be in control.

Covid lockdowns pose the same exact question as the Bystander at the switch.

It's a direct violation of the principles of universal human rights where governments around the world pulled the switch by imposing lockdowns "for our safety"—in doing so they gave themselves the authority to play God with our lives.

No police came to the Boston man's aid, nor did 911 answer the call. The victim was asserting his constitutional rights to not wear a mask all over Boston and the police knew who he was. The police also knew he was correct in his assertion of those constitutional rights and couldn't assert their authority on the victim because he was correct in said rights.

In retaliation they decided not to uphold their sworn duty to uphold the Constitution of the United States. Because the evidence was caught on video, a detective contacted the victim after the fact.

New Berry St. is the busiest street in Boston with cops at every corner except for this particular day where the victim drove around for hours looking for a cop and tried multiple times to call 911.

This is why 90% of Americans want to physically harm the police. When the average person is thinking of ways to improve the Molotov cocktail to destroy these aggressors you can clearly understand we have real trouble.

Now let's think defunding police through. People like me are ready to eradicate aggressors and thieves with violence. There are a lot of people, whom are dear to me, who cannot and it's so far from their consciousness that they would rather die than kill. Y'all are stronger than I.

So the need for police in my circumstance is only to keep the police from aggressing on me for asserting my constitutional and universal right to protect my person and property.

Most Americans are looking for a superhero to save them and there aren't any, that's fiction. They want someone to do their dirty work, but when they get to dirty, Americans want too throw them away.

The only way we can continue and correct the police problem is to have them become peace officers as they once claimed, and an understanding by everyone that the police's job is to ensure American's constitutional, universal, and unalienable rights. Protect and serve is bullshit; you are required to protect yourselves.

In Luke 22:36, Jesus said to sell a garment and get a sword. The time when you need not purse, scribe, or shoes is gone.

To have humans believe the State is going to solve their problems was a way communist factions twisted the fabric of our society, to destroy the Constitution. To be off the mark is called Sin, it's an Archer's term. The degree of sin is how far off the mark you are.

<u>The bullseye isn't defund the police, it demilitarize the police.</u> They don't need tactical weapons, tanks, or drones, they are peace officers who protect constitutional liberties. Psychological conditioning has yet again distorted you simpletons.

Is it a surprise to anyone that Philip Zelikow, the former Executive Director of the 9/11 Commission, will chair the Covid Commission? Probably not.

Zelikow, a President Bush insider who helped Bush move into the White House with the transitional team, also co-authored a book with Condoleezza Rice called "To Build a Better World." Zelikow also wrote the 2002 "The National Security Strategy To The United States of America" which laid out the case for preemptive war to invade Iraq and later Afghanistan.

Even Former Counter Terror Czar Richard Clarke said "the fix was in… could anyone have a more obvious conflict of interest than Zelikow?"

Zelikow is a war criminal. Zelikow designed the authoritarian world we now live in. Henry Kissinger was the first to lead the 9/11 commission, but it was soon exposed that the Bin Laden Family was his client. Obviously that old devil stepped down. Now it's obvious the fix is in because Zelikow is on the scene.

Under Zelikow, they made it understood that anyone who questioned his authority was a conspiracy theorist. Even though most all engineers in the world concluded that in no way could the towers imploded with their structural integrity.

Zelikow wants you to believe these crisis will be more common and we need to learn to strengthen our society. He went on to say scholars and journalists will do their job, but also there must be massive research to investigate this crisis, only that which a large scale commission can provide. Zelikow wants money and control.

When Zelikow says journalists, does he mean CNN, The New York Times, USA Today, and Facebook? Judging by his smug face I'd say yes.

They will be backed by four leading charitable foundations: Schmidt Futures (owned by Eric Schmidt), the SKOLL Foundation (owned by eBay), the Rockefeller Foundation, and Stand Together (owned by Charles Koke).

They will also work with the Event 201 participants like the Johns Hopkins Center for Health Security and the Bloomberg school for public health.

How about Event 201?

The Johns Hopkins Center for Health Security in partnership with the World Economic Forum and the Bill and Melinda Gates Foundation hosted Event 201, a high-level pandemic exercise on October 18, 2019, in New York, NY. The exercise illustrated areas where public/private partnerships will be necessary during the response to a severe pandemic in order to diminish large-scale economic and societal consequences.

"In recent years, the world has seen a growing number of epidemic events, amounting to approximately 200 events annually. These events are increasing, and they are disruptive to health, economies, and society."

"Managing these events already strains global capacity, even absent a pandemic threat. Experts agree that it is only a matter of time before one of these epidemics becomes global—a pandemic with potentially catastrophic consequences. A severe pandemic, which becomes Event 201, would require reliable cooperation among several industries, national governments, and key international institutions."

Event 201 simulates an outbreak of a novel zoonotic coronavirus transmitted from bats to pigs to people that eventually becomes efficiently transmissible from person to person, leading to a severe pandemic. The pathogen and the disease causes are modeled largely on SARS, but it is more transmissible in the community setting by people with mild symptoms.

This is to psychologically condition you to believe SARS is from animals when in fact it comes from the "Specific Adsorption Rate" of radio frequency.

This all from the Center for Health Security.

They go on to "simulate" a sick pig in Brazil makes poor people die. Starts off slow, then increases by air travel to Portugal, United States, and China. There's no drug that can help, and death becomes severe. After 18 months, 65 million people are dead. Only after 80-90% of the Globe is vaccinated does it slow, after it will only be children who become sick and die.

This all in October 2019 mind you. They knew people will become sick. Through propaganda in movies, and television shows they knew how to imbed the story line first.

What they didn't tell you is the technology that is Bill Gates' livelihood (built with communist slave labor) for income is the product that will make you sick, along with a massive amount of pollution from Africa and Mexico. The 5G technology with its millimeter wave length has caused every living being to become sick, and just a few months after the start-up of 5G. We learned about this non-ionizing wavelength earlier.

Now Bill Gates is getting divorced. Why you ask? When Enron was about to go down the tubes one of the CEOs divorced his wife and was able to sell his stock without having to disclose it to the shareholders. Disclosure would have made the stock significantly drop and he would have lost a giant portion of his income. Also he didn't go to jail like the other Enron assholes who did sell their stock without disclosure. Some tricky shit, but legal.

After they vaccinate everyone they think they can, it will slow down the ability for children to be born from said vaccine—the democide Bill Gates has worked on for so many years.

They also intended to stop air travel because we were going to suffer real pollution in 2020. They took away the lesson humans needed to learn and now it will linger on because they believe they know better. Instead of letting the child feel the hot stove they put up a gate and hired security guards to stand beside the stove. All the while the child is thinking—I wonder what the stove feels like.

You cowards prolonged our pain, and now it is you that will pay the consequence. We needed to feel the pain of dying from pollution and now it continues.

If you don't believe the Center for Health Security then they have fact checkers like USA TODAY to make sure you don't receive misinformation. I wonder if they used the New York Times polling opinion, and later fed it to Facebook.

On March 26, 2020 USA TODAY headline reads —Fact check: A Bill Gates-backed pandemic simulation did not predict Covid - 19.

The story says: *"Event 201 was a real operation, there is no evidence that it was meant to model or engineer the current COVID - 19 pandemic."* That is true, however with the lack of journalism integrity or the ability to think for themselves, they also didn't ask the *"invite only Event 201 participants"* if they knew people were going to become sick with the start-up of the 5G infrastructure. Or if the closed door meeting was most likely how to mitigate their financial liability.

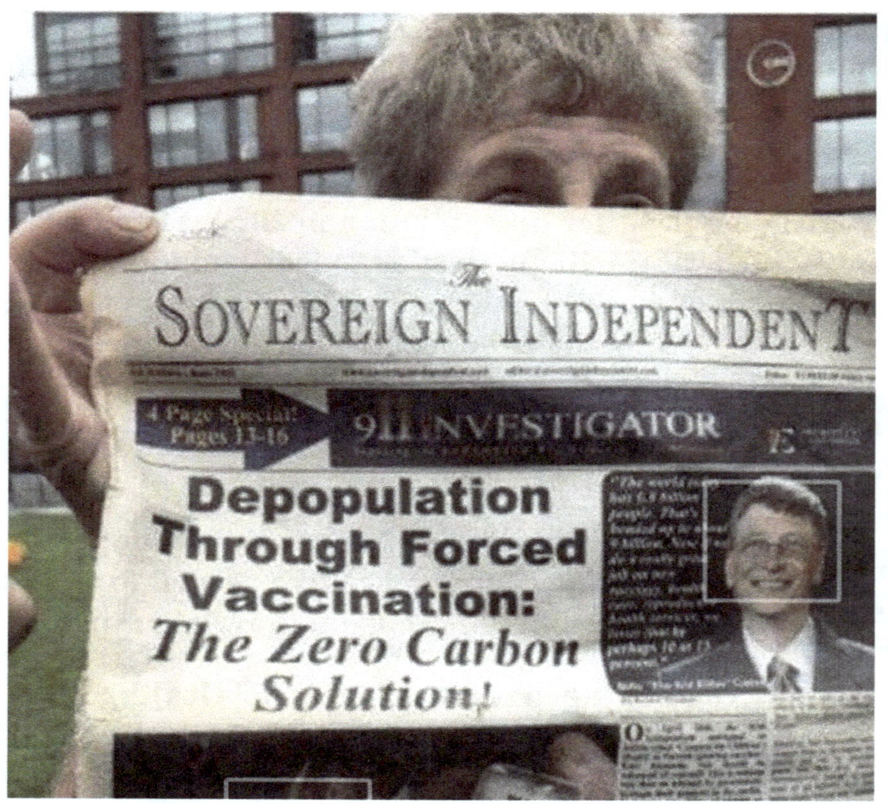

In a statement to USA TODAY, the Bill & Melinda Gates Foundation cited several partnerships and contributions to pandemic preparedness it had increased since the 2014-15 Ebola outbreak.

Y'all remember when they tried to convince us African Zombies were after us? They watched too many movies and are ill-equipped to lead anything.

Another USA TODAY Fact Check: Did Bill Gates predict the coronavirus in 2015? In a Feb. 28 article for The New England Journal of Medicine, Gates wrote, "In the past week, COVID-19 has started behaving a lot like the once-in-a-century pathogen we've been worried about. I hope it's not that bad, but we should assume it will be until we know other-wise."

America, at what point is it ok for a computer geek to give an opinion in The New England Journal of Medicine, or to give direction on viruses? Actually, he's not even a computer geek, he is a monopolist. I guess because he has lots of money he must know about everything? Maybe the The New England Journal of Medicine is easily bought and are regular ass kissers.

If Bill Gates wants to be prepared then he should. Believing he has authority over anyone else other than himself is going to get him killed. Of course people want to go to his 201 Event, because Bill Gates has more money than sense—the Communist Party and slave labor made it so. The participants know they can kiss his ass and go home with a gift basket or funding like Phillip Zelikow wants. Phillip Zelikow's greed and deception makes it clear that he is an ene-my of all living beings.

It's most likely because all people in the indus-try already knew the millimeter wavelength (5G) was going to kill. They would do anything to direct your consciousness elsewhere.

When you hear the words Nuclear Energy you think wow, this is some advanced technology. Only the most brilliant minds could come up with this, right?

I'm here to tell you that nuclear energy is an over designed steam engine. The whole process is de-signed to boil water to turn a fucking squirrel cage to create energy. We'd be better off getting some squirrels and nuts.

The detriment far out weighs the benefit. There are miles of underground tunnels in the high desert of the Northwest that are filled with double lined stainless steel containers that contain massive amounts of nuclear waste. It's purpose is storage time in attempt to allow it to age to its half life. Scientist claim its half life is less harmful, and I'm sure that it is half as dangerous. I didn't need a scientist to understand that.

However, an earthquake, flood, or the wrong people in charge of the hazardous material could be very detrimental to a third of America's water supply for hundreds of years. North Austin has the same troubles—military grade poison. I believe it's intentionally positioned for an attack.

We can understand this from Chernobyl and Fukushima Daiichi nuclear power plants. One had a meltdown the other had an earthquake.

Fukushima Daiichi nuclear power plant has wrecked a portion of the Pacific Ocean and continues even now. As a friend of mine said, "If there wasn't a Godzilla already, there is one now."

We must stop allowing television and movies to cloud our better judgement. We are still in the Dark Ages when it comes to energy. People still think NASA and Space X are futuristic, when they both need combustion to create movement. This is some caveman shit. If you have to create an explosion, to turn a piston, to turn a wheel, then you are a fucking caveman.

Even electric vehicles need combustion from coal, gas (natural and diesel), and nuclear energy. Believing you are saving the world because you have an electric vehicle is asinine. We are still receiving energy from combustion. That's how we mostly create electricity.

White people are eager to get to space and are ready to discard the earth to the minorities. To trash the earth and leave it—like a neighborhood that is old and outdated. They focus on space when we know nothing about our ocean, the science of hearing, and the Austin, Texas power grid ain't worth shit.

NASA would rather look into the Abyss than to look in our own back yard; it's called procrastination and greed.

"The extension of life beyond Earth is the most important thing we can do as a species." says Elon Musk. I'll bet Musk doesn't know how many children died of starvation in Jefferson County, Texas alone. That's because "Tony Stark" is too busy exploding rockets in South Texas and destroying our water supply.

Come to think of it, we have a massive amount of industrial metallic ash coming from around Corpus Christi that spreads over Texas. I'll bet he has a free pass from Texas Commission on Environmental Quality; TCEQ has the love of money.

People's fear of death has clouded their better judgement. It's ok if we die, we don't need to upload our consciousness into a computer avatar, or fly off to another planet that has nothing on it. You are going to die and there is nothing you can do about it.

This Locust trait must be weeded out and eradicated. I say no more playing space man until you clean your room, mow the lawn, take the trash out, walk the dog, do your homework, and a bunch of other shit.

Our priorities are clean food, water, and air that is available to all living beings effortlessly. Healthy mind, body, and soul to everyone. Freedom for all without aggressing on anyone. Living within the real limitations the Earth provides. Until then, no more space man.

The House of Representatives approved gender-neutral language in new rules for congress like— Mother and Father will be Parents, Aunt and Uncle will be Parent's Sibling, and Niece and Nephew will be Sibling's children. The House of Representatives are now the thought police.

They sell us diversity as LGBTQ+ and ignore indigenous culture as savage.

It is obvious to me that the House of Representatives are mentally ill and must be rectified. You need to be ready to lynch your favorite politician at all times. Idolatry will damage your existence.

People identify themselves as inclusive because they finally accept queers and black people. However they turned away from everyone else that doesn't.

They are only inclusive to the prescribed inclusiveness.

Ludwig Von Moses pointed out the paradoxical position of "Progressives" who *"insist the consumers are too ignorant or incompetent to buy products intelligently, while at the same time touting the virtues of democracy, where the same people vote for politicians whom they do not know and for policies they hardly understand."*

"We have the ability to comprehend a products value after purchasing it and realizing we don't like it. But when we purchase a politician, the purchase lacks the direct test of success and failure, not to politicians whose measure have the best chance of success, but those with the ability to sell their propaganda."—Progressives.

Why is there such a mechanical governor on intelligence—the House of Representatives buckled to the deconstruction of the family unit in disguised as inclusiveness? These claimed inclusive people say they believe in Science, while clinging to mysticism of gender. This is like Qanon—it is a cult with customs, rituals, and belief systems in conflict with reality.

If you neuter a dog does it make it female? These "Woke People" put to sleep what they left behind. I fear it'll take too long before these Wokes finally climb from the proverbial rabbit hole.

It froze in February of 2021, and the power grid failed because it's been in decline for a long time. Well over 250 thousand people froze their asses off. The

Governor of Texas could have salted and sanded the roads so people could move around, get groceries, help each other.

Greg Abbott and everyone who is in love with their precious oils wanted everyone to suffer for even considering leaving fossil fuels—like an abusive husband. To condition your mind to beg for even a dumpster fire to create energy. There's a lot of things that could have been done, but he's only a lawyer, he only knows how to be a lawyer.

Someone said the Texas power grid is better than California's power grid because it only failed once. That's because California is over populated and Texas is well past the beginning stages of over population. All big cities in Texas are over populated to a detrimental point. Here's a good way to condition your mind; let you freeze in the cold so you'll beg for coal fire.

You are drunk with ignorance. It has overpowered you, and now you are vomiting it up. Empty yourself of darkness and you will be filled with light.

XXIV

1923 - US Air Service created a smoke screen for military purposes.

1932 - GE using Vincent Schaefer method and seeded the first Cloud.

1946 - Project CIRRUS impregnated (using Schaefer method) a hurricane close to Jacksonville, Florida. The storm moved inland and damaged Savanna, Georgia leaving 105,000 people homeless and killed many.

1952 - Project CUMULUS or Witch Doctor, on August 15, a storm killed 35 people and 400 homeless in the seaside village of Lynmouth, Great Britain. People who worked the project said they did seed the clouds that washed the village to the sea.

1953 - Project SKYFIRE, weather modification with silver iodide.

1961- Project SKYWATER, Congress directed the Bureau of Reclamation and are accomplices.

1962 / 71 - Operation RANCHHAND, the spraying of Agent Orange destroyed vegetation and food from the Vietnamese, causing birth defects in Americans and Vietnamese. High levels of dioxin is still in the soil and water. Agent Orange produced by DOW Chemical.

1962 - Project STORMFURY, the use of silver iodide.

This same year (1962) LBJ gave a speech at Southwest Texas University and explained that we did have the capability to control the weather and in his words "he who controls the weather, controls the world." Just what you can expect from someone like him.

1966-72 - Project POPEYE, the use of weather as warfare, LBJ liked the idea but was ignorant to the final outcome.

1976 - America started having strange unexplained blackouts in worldwide communications. Russians developed and covertly implemented the ability to send ELF radio waves (low frequency waves) into the atmosphere and controlled the jet stream, the US developed HAARP to do the same and both using ELF Waves. It made a clicking sound the Pentagon referred to as the Russian woodpecker.

HAARP, (High - frequency Active Auroral Research Program) - a joint effort with the Airforce and Navy and with cooperation with academic institutions. The projects consist of 80 antenna 70 ft tall linked together and steerable. HAARP can produce millions of watts of ELF into the atmosphere and is one of many ELF transmitter stations on earth. Working in tandem, the transmitters alter the weather and the waves get reflected back down and pass through the earth, ocean, and all living beings.

1980 - It was reported and the thought was jet engine exhaust or contrails were making it more cloudy especially in the mid-west and studies suggest erratic weather occurred after and was proven to reduce crop production.

1982 - a Pentagon researcher found the Russian wood pecker was creating an artificial ionization in the upper atmosphere and bending the jet stream, altering weather.

The United Nations IPCC (Intergovernmental Panel on Climate Change) suggest global warming is man made and the panel consists of non-scientist, political hacks, and order followers. Climate scientists need the problem to get funding and have a vested interest for panic, it's a giant industry.

2013 - Al Gore appeared on the Ellen DeGeneres show and explained sulfur dioxide was released into the atmosphere to blockout the sun's heat in an attempt to stop the warming of the Earth from pollution by using pollution. It's called Solar Radiation Management.

That same year on the CBS morning show Nora O'Donnell interviewed physicist Michio Kaku who explained the use of trillion watt lasers to precipitate rain clouds and bring down lighting bolts down the beam. Only if there is water vapor in the air could it be condensed.

Geo-engineer David Keith claims we emit 50 million tons of sulfuric acid "spray pollution" into the sky to help with global warming so we don't have to cut emissions and verifies it kills millions of people every year. The idea stems back to LBJ and they are selling us the solution and problem. The oil industry turned our atmosphere into a toxic waste dump. It's cheap to spray chemicals into the atmosphere.

2016 - CIA Director John Brennan explained on C-SPAN at the Council on Foreign Relations that Stratospheric Aerosol Injection (SAI) would give the world economy time to move from fossil fuels.

We have arrived at the point in your read to understand the millstone tied around humans' necks. The god of the world economy desires to survive and will destroy all to do so. This is why millions of living being have been tortured, enslaved, and murdered. This is the pivotal point of change for all living beings and those who are to come. The powers and principalities are desperate to survive, but the banking cartel will allow it not.

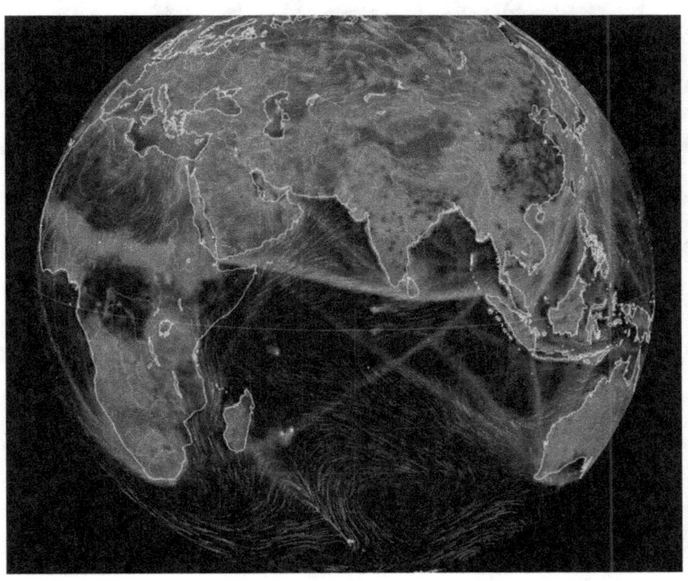

2011 - Ted L. Gunderson, former Chief Agent in charge of FBI in Los Angeles, Dallas, and Memphis says the UN is geocoding humans and animals with the "death dump" of chemicals into our atmosphere. Gunderson personally witnessed the plains that are loaded with the hazardous material in Lincoln, Nebraska—the plains that spray hazardous material over our unknowing population. All members of the UN and representatives have grown tired of living, and are suicidal. We must deliver them all from their pain.

On January 1, 3, 28, and on February 1, 2021, I made a post on Instagram, under the handle UNCLE_SWEETHEART, showing sulfur dioxide from Canada fueling the massive winter vortex and raising Covid numbers on the East coast .

On February 16, 2021, I also showed the sulfur dioxide aerosol thickness from Canada to Mexico that caused the freeze in Texas that caused million of power outages, deaths, and Covid numbers to rise.

On August 18, 2021, I made a post on Instagram, under the handle UNCLE_SWEETHEART, showing a newly forming hurricane in the Atlantic and one not yet formed in the Caribbean.

By August 20th, it was clear to see, through real time satellite imagery from <u>earth.nullschool.net</u>, that the hurricane on the Eastern seaboard was being fueled by ground level background ozone and fine metallic soot. Also a hurricane formerly not formed in the Caribbean was traveling into Mexico and was accelerating from the center of the pollution.

On August 21, 2021, I showed on Instagram, a massive amount of ground level background ozone and fine metallic soot coming from the continent of Africa. The image also shows the pollution fueling the hurricane as it traveled up the Eastern seaboard.

New England has a constant source of pollution from the continent of Africa. The sulfur dioxide surface mass from steel country (Pennsylvania region) and coal plants to create electricity causes extreme winter weather. Hurricanes and Polar Vortex are all caused by the disruption of our natural breathing air with weather modification and the use of pollution as a conductor for weather modification.

On August 25, 2021, a newly formed distur-
bance, Ida, developing between Central and South
America—fueled by pollution from Africa.

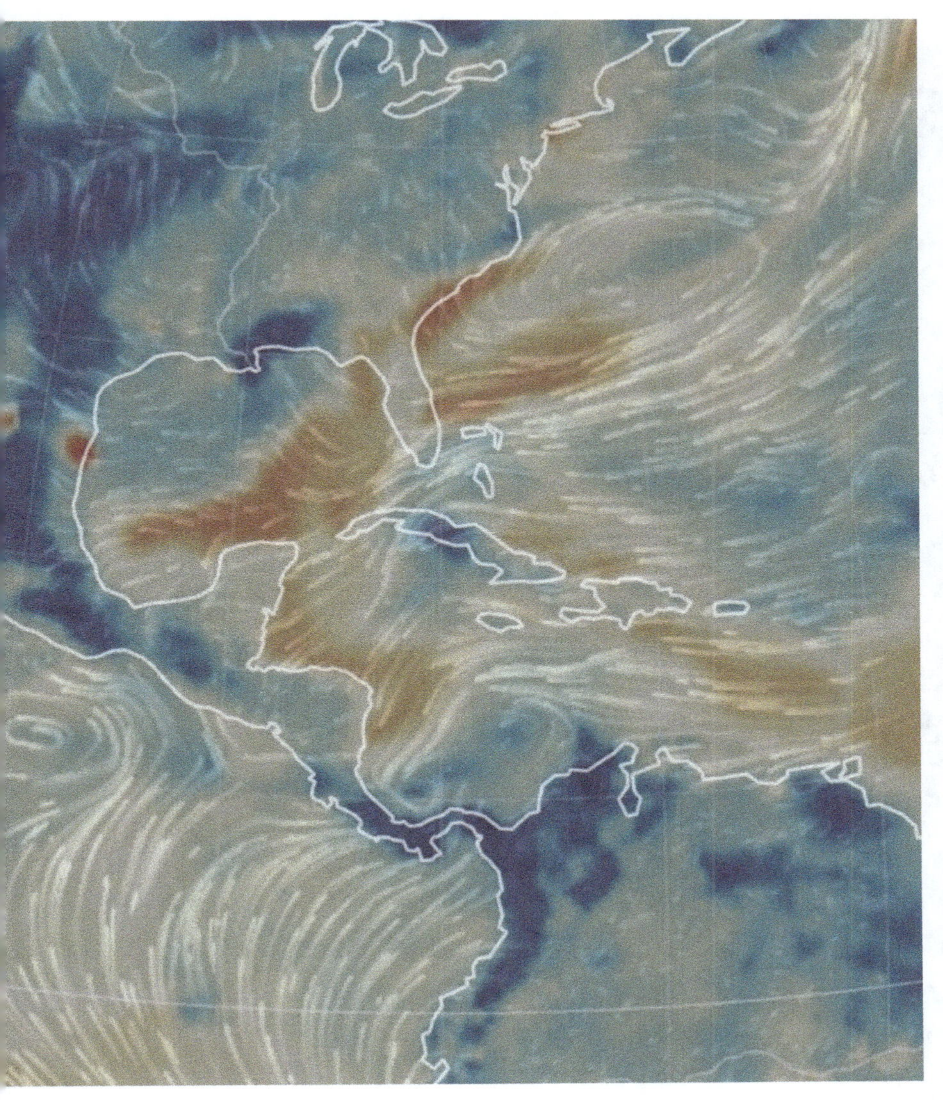

August 27, 2021, Hurricane Ida formed on the the south coast of Cuba and strengthened from the pollution between Cuba and Florida. The pollution drew hurricane Ida in the center of the path of high concentration.

August 28, 2021, Hurricane Ida traveled and consumed the pollution. By August 29th, Ida made land fall in the direct path of the highest concentration of ground level background ozone and fine metallic soot. We know it's not just dust from Africa because TCEQ showed on their air quality forecast that it was Ozone/PM2.5. The extra oxygen molecule in the ozone strengthens the storm as oxygen in an oxygen acetylene cutting torch would.

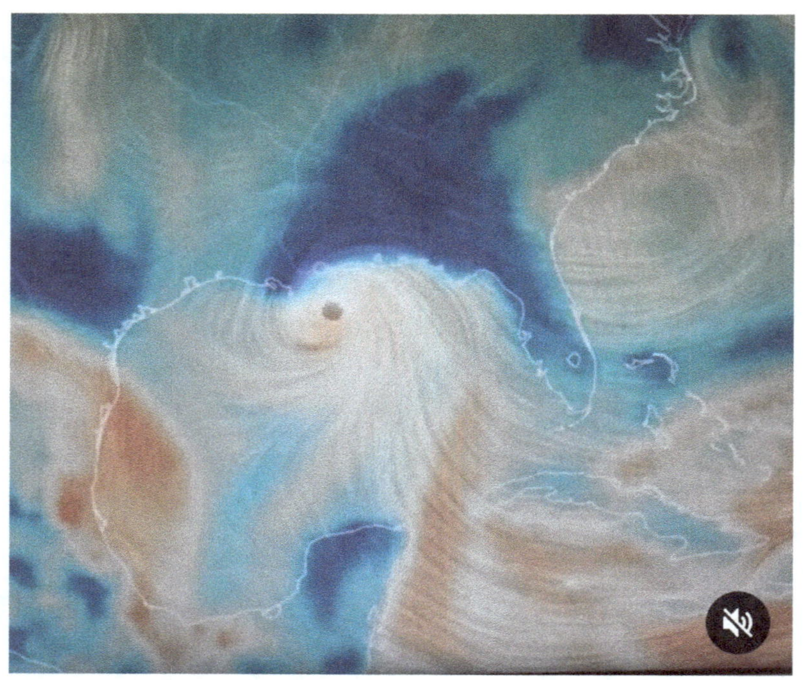

This is happening every year for many years and many years to come if we do nothing about the pollution from Africa.

The loss of homes, businesses, and livelihood equal more money than the owners and operators of the pollution lives are valued at. The owners and operators of this deadly and dangerous pollution are in violation of Universal Law and must be punished with extreme measures. They are mass murdering everyone, we must return the favor. This is warfare by both the toxic air and the weather modification it creates. Believe and act.

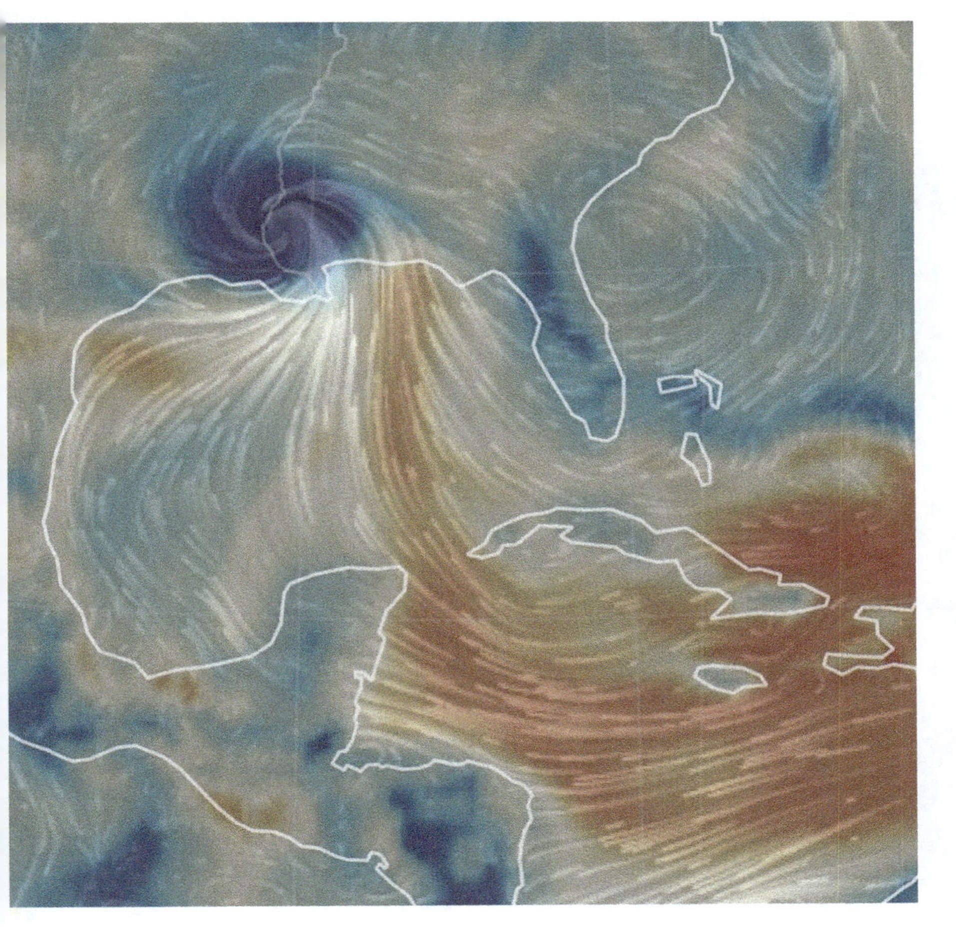

First Nations

XXV

When Europeans came to this land, they had the Bible and taught the natives how to pray with their eyes closed. When they open their eyes the natives had the Bible and the Europeans had the land.

The Europeans would have not survived in Jamestown had it not been for the help of the Powhatan in 1609. The Europeans spent most of their time digging for gold because propaganda convinced them the New World was dripping with it.

They didn't prepare shelter, clean water, or food supply. They relied on the supplies they brought which ran out. The Europeans were digging up dead human bodies to cook and later were charged with the crime of cannibalism. The Powhatans felt sorry for these ignorant beings and took them into their homes, feeding, clothing, and keeping them warm through the winters.

To repay the Powhatan's good deed, the tribe was violently decimated by the English colonists in 1646, and suffered measles and small pox. They didn't have disease until the English brought it to them. They should have annihilated the English before they could get a foothold, but that was too far from their consciousness at the time. They only knew to help living beings, they didn't understand these demons.

In 1681, The Penn family made a treaty with the Lenape native, which included the Delaware natives. William Penn patted himself on the back for his fair dealing with the natives. The truth is the Penn

Family were nothing more than Pharisees and demons.

This treaty was known as the "Walking Treaty." Two natives and two who represented for the Penn family would walk for a day, on the next day walk in another direction until a square was made by walking.

The first day they walked until dark and set up camp. The next day, different Penn family representatives relieved the first delegates, only this time they ran as fast and far as they could. Then again on the next day more Penn family delegates relieved the exhausted runners and ran again in an attempt to gain more land by running; this is how demons work. Even though the dishonorable Penn family was obviously treacherous, the natives let them have it anyway. This was a mistake, you can give not a demon an inch.

Even after all the deception, the demon's offspring (the two Penn sons) still claimed more land and sold it to English colonists while the natives inhabited it. That was an evil way of having human beings attack each other while the true perpetrators benefited monetarily. The colonist paid rightfully for land and the seller was fraudulent. It's up to the new colonist to fight off the savages, while the Penn family gets two birds with one stone.

The natives didn't believe or understand what they saw. "Who are these new invaders? They have not right to be on our land, we must defend it," thought the natives. When demons show themselves, believe them. They should have killed the Penn family at their arrival.

In 1804, Sacagawea was a 16-year-old Lemhi Shoshone woman who was kidnapped and sold to a French Canadian trapper named Toussaint Charbonneau, whom also had purchased another native named Otter Woman.

Toussaint Charbonneau and Sacagawea were commissioned to join the expedition because Sacagawea could communicate with the surrounding natives. Lewis and Clark also utilized the appearance of having a woman and child as friendly, especially a native woman.

Sacagawea saved their journals and documents from a capsized boat and they named the river after her, mighty white of them. They traded for horses with the shoshone and discovered the chief was her brother—Cameahwait.

The Shoshone agreed to barter horses and provide guides over the Rocky Mountains because of Sacagawea. They almost starved to death, eating tallow candles to survive. When they got over to the other side of the Rockies, Sacagawea found and cooked camas roots to keep them alive.

Sacagawea gave up her beaded belt for an Otter robe for Thomas Jefferson. She also showed them a trail her people were familiar with later called Gibbon's pass and Bozeman pass which developed into the Northern Pacific Railway to cross the continental divide. Sacagawea did all this with a baby on her hip.

For her good deeds William Clark took and adopted her two children. Did she entrust her son Jean Baptiste Charbonneau to Clark, or maybe she didn't have a choice because she was a slave wife to Toussaint Charbonneau? Sacagawea longed to be with her people and they took her children to give a compulsory education. It appeared that Sacagawea died of a broken heart.

Sacagawea's payment for making the expedition susessful, saving everyone's lives, and helping the future murderers navigate the continent, was the theft of her children and perpetual servitude.

But wait, the gift just keeps on giving.

The Shoshone allowed fur trappers to hunt along side of them. The fur trappers recommended Brigham Young and the Mormons to move into Shoshone territory because the food source was plentiful. But, plentiful for the inhabitants only. They should have cut the head off that snake immediately.

Around 1847, the Mormons move into Seuhubeogoi or Willow Valley and use up the resources like hunting game and foraging plants. They over populate the valley and the Shoshone starve, having to eat livestock owned by the Mormon settlers.

Despite the policy of "feed the native instead of fighting them," the settlers were consuming significant amounts of food resources and taking over areas that pushed the Shoshone increasingly into areas of marginal food production. Tensions were high and the settlers looked at the natives as animals.

More and more warfare resulted in the Bear River Massacre in 1863 when the US military killed an estimated 410 Shoshone at their winter encampment. The military killed men, women, and children. Only a demon could order and perform this action, a demon like Abraham Lincoln. How do I know Lincoln is a demon? Because he is still revered as benevolent.

December 26, 1862, Abraham Lincoln hung 38 Dakota Sioux natives in Mankato, Minnesota. The day after Christmas this revered demon ordered this, just one week before the Emancipation Proclamation. The Sioux were angered from broken treaties, refused to be placed on reservations, and were being forced off their lands.

A witness reported them saying, "Hear me my people, today is not a day of defeat, it is a day of victory. For we have made our peace with our creator, and now go to be with him forever. Do not mourn for us, rejoice with us, for it is a good day to die." That's more faith than most all Christians combined. What kind of demon orders this action? Native Americans didn't understand that treaty meant a signature for your later theft and death.

Each of them shouted out their names in their native tongue, "I'm Here! I'm Here!"

The Emancipation Proclamation had nothing to do with freeing black slaves, only to add trigger fingers because the Union Army was falling. We know this to be true because slaves were still bound in Union held states long after the Emancipation

Proclamation. If you don't believe then look at the National Archives.gov to see, second paragraph, second sentence.

Booth killed Lincoln, which is strange to me because Democrats aren't very good shots. It was point blank so it didn't really count.

There were approximately 5 million African slaves killed, and between 80 and 100 million natives killed. It's the least discussed genocide in world history. Millions of indigenous people had been living on this continent for millions of years. Their nature has allowed for the deception and intrusion.

In no way did Jacob, who later changed his name to Israel, father the native people. In no way did the House of David have anything to do with native people. In no way are the Native American related to the Twelve Tribes of Israel (AKA Jacob whom saw a ladder ascending into the sky). Even I was foolish enough at times to believe.

The Caddo are the ancient man to most all southern tribes including the Mayans. They may be the grandfather of more than southern tribes. These tribes were in America before Moses' family (a scorned priestess family whom left Sumer before the fall) moved to Turkey.

The natives have had advanced cultures and civilizations for millions of years. The Caddo is the grandfather to the Mayans who built giant pyramids and could understand universal time and space. They have already mapped out the solar systems and understood energy. Somewhere along the line a wiping of memory caused a time of forgetfulness.

Some believe America is the land of immigrants, on the contrary—America is the land of the Caddo and more ancient beings. It wasn't until around 1880 when the natives were even considered to be human, even though they were treated like animals long after.

Standing Bear or Mantcunanjin, a Ponca chief who spoke truth within this universe, forced the U.S. District Court in Omaha to acknowledge native Americans as "persons within the meaning of the law." and also have the right of Habeas Corpus. Standing Bear was the first Native American to be "granted" civil rights under American law.

That's mighty white of the court to grant something humans were born with.

By 1862, white settlers were settling where the Ponca summer corn fields had been for generations. The U.S. government was going to relocate the Ponca and indeed showed them new land. The Ponca weren't satisfied with the options so they abandoned the search with the Bureau of Indian Affairs. The Ponca did sign a document stating they were moving to said land, but the supposed misunderstanding and translation caused the deal to fall. The result was the Ponca were forced to move to Oklahoma.

The Ponca arrived in Oklahoma too late to plant crops that year, and the government failed to provide them with the farming equipment it had promised as part of the deal. They also promised clothing and shelter as well, instead they turned them out like a coon caught in a live trap to fend for themselves.

In 1878 they moved 150 miles west to the Salt Fork of the Arkansas River, which is south of present-day Ponca City, Oklahoma.

By spring, one third of the tribe died of starvation and malaria including Standing Bear's eldest son, Bear Shield. Standing Bear had promised to bury his son by the Niobrara River homesite. Standing Bear traveled north with about 30 followers.

Word got out to the Secretary of Interior and he ordered Brigadier General George Crook to arrest the renegade Indians.

The strange part is Brigadier General George Crook is the person who initiated the change of tide. He told the story to Thomas Tibbles, the editor of the Omaha Daily Herald who publicized the Ponca story. The chief attorney of the Union Pacific Railroad, Andrew Poppleton, and attorney John Webster offered pro bono services.

The case is called *United States ex rel. Standing Bear v. Crook.* General Crook was named as the formal defendant because he was holding the Ponca under color of law.

As the trial ended, Standing Bear had an opportunity to speak on his own behalf. The account was told that Standing Bear arose and, only half facing the court, held out his right hand and stood there motionless "so long that the stillness of death which had settled down on the audience became almost unbearable." Finally, he looked up at the judge and spoke:

"That hand is not the color of yours, but if I prick it, the blood will flow, and I shall feel pain. The blood is of the same color as yours. God made me, and I am a man. I never committed any crime. If I had, I would not stand here to make a defense. I would suffer the punishment and make no complaint.

I seem to be standing on a high bank of a great river, with my wife and little girl at my side. I cannot cross the river, and impassable cliffs arise behind me. I hear the noise of great waters; I look, and see a flood coming. The waters rise to our feet, and then to our knees. My little girl stretches her hands toward me and says, 'Save me.' I stand where no member of my race ever stood before.

There is no tradition to guide me. The chiefs who preceded me knew nothing of the circumstances that surround me. I hear only my little girl say, save me. In despair I look toward the cliffs behind me, and I seem to see a dim trail that may lead to a way of life. But no Indian ever passed over that trail. It looks to be impassable. I make the attempt.

I take my child by the hand, and my wife follows after me. Our hands and our feet are torn by the sharp rocks, and our trail is marked by our blood. At last I see a rift in the rocks. A little way beyond there are green prairies. The swift-running water, the Niobrara, pours down between the green hills. There are the graves of my fathers. There again we will pitch our teepee and build our fires. I see the light of the world and of liberty just ahead."

Taking a long pause again, Standing Bear finally turned toward the judge "with such a look of pathos and suffering on his face that none who saw it will forget it." He then says:

"But in the center of the path there stands a man. Behind him I see soldiers in number like the leaves of the trees. If that man gives me the permission, I may pass on to life and liberty. If he refuses, I must go back and sink beneath the flood. You are that man."

It is said by Thomas Tibbles, the court reporter, that tears ran down the judge's face and that General Crook buried his face in his hands— and yet the court's ruling was based not on sympathy but on the Constitution and the law.

On May 12, 1879, Judge Elmer S. Dundy ruled that "an Indian is a person" within the meaning of *habeas corpus*. He stated that the federal government had failed to show a basis under law for the Ponca's arrest and captivity.

It was a landmark case, recognizing that an Indian is a "person" under the law and entitled to its rights and protection. "The right of expatriation is a natural, inherent and inalienable right and extends to the Indian as well as to the more fortunate white race," the judge concluded.

The army immediately freed Standing Bear and the other Poncas.

The white man needed a judge to tell them a human being has rights given to them by a higher power, at birth, that are unalienable. They can be administered not or taken away, they just are.

I could fill another book on how the Spanish demonized natives. The Spanish and their priests were some of the worst. I think they were sent to the New World to get them away from the rest of society. The Spanish Catholic had been under Satanic control long before the idea of traveling to the New World ever existed. They too will learn the consequences of using God as an excuse to rape, murder, and steal.

In 1680, during the Pueblo Revolt in New Mexico and Arizona, Juan de Oñate enslaved and killed hundreds of natives. Oñate sentenced all men 25 or older to have a foot cut off. The Pueblo were force to provide tribute to the colonist in forms of labor, ground corn, and textiles.

They forced natives into labor camps called "encomiendas" while restricting them from fertile farmland and access to water.

They established theocracies in villages to control the Pueblo, and gave an appearance of Catholicism so Spain wouldn't abandoned the province.

They outlawed the Kachina dances and ordered the missionaries to seize and burn their masks, prayer sticks, and effigies. They forbade the use of entheogenic drugs such as cannabis and psilocybin, naturally occurring substances.

People like Nicolas de Aguilar defended the natives right, but they banished him in the Mexican inquisition.

The Spanish used God to justify their actions—all through Mexico, Central, and South America, raping, killing, and conditioning their consciousness. It was a willful destruction of one human by another.

Tecumseh asked, "What kind of people is this for us to trust?" He went on to say, "When Jesus Christ came upon the earth, you killed him, the son of your own God, you nailed him up! You thought he was dead, but you were mistaken. And only after you thought you killed him did you worship him, and started killing those who would not worship him. What kind of a people is this for us to trust?"

On and on, the theft of land and the willful destruction of a human by another. Why must these demons do this?

Because of Blood Quantum.

The bloodline of the Native Americans keep the demons, Sanhedrin, Chief Priests, and Pharisees from completely owning and controlling everything in existence. If the tribes no longer exist, then the demons own the land and can rape it until there is nothing. Not all of us are pure blood, but some have some blood. Which brings me to the hundreds of missing and murdered indigenous women in present day.

Women and girls from 41 tribal nations are represented in these missing and murdered cases. Nearly 100 victims are ages of 18 and under and 59% of the missing and murdered cases are on reservations. Witnesses identified 1 in 4 alleged perpetrators, most were acquitted or never charged.

The KXL pipeline was proposed in the year 2000 which 69% of the cases occurred during this time and 1 in 5 were near the proposed pipeline.

Law enforcement and the Department of Justice classified one third of the cases as accidental, exposure, natural causes, overdose, or suicide without an adequate investigation.

These demons want to stop the natural native bloodline to break the illegal treaty with the natural owners of the land. Even a demon has to pay for breaking a contract and has to plea their case to God to justify it. These demons take kindness for weakness and are in violation of the contract. They are also guilty of genocide, murder, theft, and more. Those are stiff charges in the Universe.

Femicide - the killing of women.

It's happening all over the world. A woman in Mexico was raped, tortured, and skinned like a deer. The same with the Aboriginals of Australia, in Central and South America, all over the world.

Even the tribal people of England suffered William the Conqueror, "a French bastard who landed with an armed Banditti and established himself king of England against the consent of the natives." This is the monarchy's origin.

On the 27th of May 2021, the remains of 215 children (some as young as three years old) were found on the site of the former Kamloops Indian Residential School. The residential schools were operated by the Catholic churches from around the 1840s to 1990s where native children were forced to convert and attend.

This isn't the only residential school in Canada or America, there are many. Ottawa says it's not liable for cultural damages caused by Kamloops residential school, reports Canadian Broadcast News.

The schools violently force the English language on the children, and ban native cultural practices. It is said that over 4,000 students died while attending the schools and the children's families thought they were missing.

Reminds me of when Queen Elisabeth II was accused of stealing 50 native children. It is said the queen took them on a picnic and they never returned. This accusation has come from many continents with indigenous cultures. The truth in the accusation is unknown, as many lies as we suffer, who can tell. Everyone is innocent until proven guilty. If evidence arrive—we will feast on them all.

George Manuel attended the school in the 1920s and said it was cold and unsanitary. It was forced assimilation and the children were taken from their families. Only demons with great darkness can do this. There was physical, emotional, and sexual abuse and forced labor. So far the death count of native children in Canadian Residential Assimilation Institutes is now 7,310—that count is for Canada alone.

This darkness is afraid of the Native and should be. The Darkness has been exposed and the fall of these low vibrational beings is near. The Nothingness of these dark forces are in violation of universal law which is now invoked.

I say we need to start hunting these killing demons. You have to find good hunting ground, trail them, make a food plot, find high ground with a clear shot, wait patiently.

Bastrop County gives $5 an ear or tail for wild hogs, I say it's at least worth that. With evidence, $5 an ear or tail.

Social media like Instagram and Facebook takes down posts that address native issues like the missing and murdered indigenous women and girls or MMIWG2S. Instagram and Facebook are always on the wrong side if that gives you any indication.

I call onto you—brothers and sisters of the Americas and the world—from the jungles to the desert, and the mountains to the seas. Hear me and understand what you read. Join me in some way to stop this destruction and shine light into darkness.

Everybody wants to be a warrior until it's time to do some warrior shit. You Cross-fit and Spartan race up until they exploit your fears, then you're out.

If you don't know what to do then hear me for those who can hear, no one has the right to aggress on another's person or property, you have unalienable right given to you at your birth. These rights can not be administered or taken away because they were given to you from a higher source at birth. If you don't believe that, then perish.

You must be a little homicidal to protect your person and property and have the understanding in your heart so that you can act accordingly. Remember, it's not who you kill, but who you let live.

The distinction between being homicidal and psychopathic is simple, one cares about living beings and the other only wants to win at all cost.

Sam Houston ran away from home when he was a teenager to live with the Cherokee, and was a citizen of the Cherokee nation known as the Raven. He also fought with Andrew Jackson against the Creek and was almost killed in the Battle of Horseshoe Bend. Houston also led the removal of the Cherokee from Tennessee to Arkansas, a white man living between two worlds.

Even I can find fault with Houston for removing his people from their native lands and fighting the Creek. However, he brought many natives to Texas to enrich our culture. Not only to enrich Texas, but to also give Texans an exit plan if necessary.

Texas is its own country that can secede from the Union at any time. We don't need permission from anyone to do so. We also have the opportunity to join the Native Nations if they would have us. All states that have Native Nations have the opportunity to disregard the U.S. and join a land within the land. This was one more saving grace, a way to grow from within outwardly.

Houston brought natives into Texas to better us, and in just a few years, the second president of Texas, Mirabeau Buonaparte Lamar an attorney from Georgia, brutally ran a large number of them out just a few weeks after taking office.

Lamar wanted to remove the Cherokee, Comanche, Delaware, and all other natives even if they had to be destroyed. Lamar wanted total extinction of the tribes to make the land available to whites.

Lamar ordered an attack which is referred to as The Battle of the Neches where Houston's friend Chief Bowles or Di'wali was killed. Di'wali, Gatunwali, Big Mush and other chiefs asked for time to gather their crops but Lamar wouldn't wait.

The Texians (Col. Edward Burleson, David G. Burnet, and Albert Sidney Johnston) advanced on the 18 Cherokee that were fit to fight, the other natives were old, women, and children. They fought all day and were backed into a ravine. The Cherokee killed five Texians and wounded twenty-seven. Hundreds of Cherokee and Delaware were killed—old people, women, and children.

Di'wali, who wore a military hat and sword given to him by Houston, fought bravely encouraging his eighteen men to fight.

Knowing the end was near, Di'wali turned his horse and charged the advancing Texians. The Texian forces shot Di'wali's horse and then shot him in the thigh and the back. Di'wali arose himself to a sitting position on the ground and sang a war song when Capt. Robert W. Smith shot him in the head. Smith then took the sword from Di'wali's body and swaths of skin from his arm as a souvenir.

John H. Reagan recalls, "I had witnessed his dignity and manliness in council, his devotion to his tribe in sustaining their decision for war against his judgment, and his courage in battle."

Di'wali was left on the battlefield according to tradition. Sam Houston was killing mad with Lamar, Di'wali was his friend.

Burnet and Johnston were cited "for active exertions on the field" and "having behaved in such a manner as reflects great credit upon themselves." It was 500 Texians against 18 Cherokee, they were pretty tough up against women and children. Or maybe they just liked how evil could take them over.

Lamar left office with the Republic of Texas $7 million in debt compared to the $1.4 million on his inaguration. The majority of the debt was accrued from carrying out his policies.

Houston led the attack that defeated the Mexican army, and with courage developed the start of Texas. It makes you wonder why anyone would want to accomplish wonderful things with great triumph, just to have some asshole destroy it just a few years later. You can make the greatest discovery that relieves human suffering, then you die as everyone will, and the next generation destroys it in minutes.

What dark, controlling force made the founders and signers of the Republic of Texas kill themselves? Some shot themselves, George Childress, the author of the Texas Declaration of Independence, gutted himself in Galveston.

Lamar won the presidency by a landslide because his opponents were dead. The first candidate, Peter W. Grayson, committed suicide, the second candidate, James Collinsworth, committed suicide by walking off the edge of the ship in Galveston, never to be seen again. Collinsworth stood up like a zombie and walked off the edge.

All this during the campaign process, all running on behalf of Houston because he couldn't run again. Robert Wilson announced his candidacy too late and Lamar won by a landslide.

The dark forces that wanted the natives destroyed had their candidate and would work tirelessly to ensure it. Southeast Texas is built on people like Mirabeau Buonaparte Lamar, and the residents are known as people of hyper-aggression and sexuality. They are easily convinced to destroy for profit, and have a psychopathic up-bringing.

Southeast Texas is known as the "Asshole of Texas," and some refer to the population (unbeknownst to the residents) as "Beaumonsters" as a play on words of truth.

Many think Southeast Texas is cursed. The first oil boom and refinery started here—the very thing that chokes our lives away now. If we are cursed, it's because of the broken Universal Laws at the beginning of conception, and the willful destruction of the natives and the rightful owners of the land. Not just in Texas, but all over the world.

Reparation - the making of amends for a wrong one has done, the act of repairing something.

We must make amends and repair the wrong that has been done or continue the suffering.

XXVI

In a farewell address in 1961 Dwight D. Eisenhower explained all to come and we were conditioned to understood not—and here are our grave errors.

We let the weight of the combination of acquisition of unwarranted influence, whether sought or unsought, by the military-industrial complex and the rise of misplaced power to exist and persist that destroyed our liberties and the democratic processes.

We fail to comprehend its grave implications. Our toil, resources, and livelihood were all involved; even the very structure of our society.

The solitary inventor has been over shadowed by task forces of scientists in laboratories and testing fields. We were not alert to the equal and opposite danger that public policy that became the captive of a scientific-technological elite.

We have become a community of dreadful fear and hate, and not a proud confederation of mutual trust and respect.

The very thing that persisted and conditioned our society attempted to complete what you will now read from Dwight D. Eisenhower:

"Throughout America's adventure in free government, our basic purposes have been to keep the peace; to foster progress in human achievement, and to enhance liberty, dignity and integrity among people and among nations. To strive for less would be unworthy of a free and religious people. Any failure traceable to arrogance, or our lack of comprehension or readiness to sacrifice would inflict upon us grievous hurt both at home and abroad.

Progress toward these noble goals is persistently threatened by the conflict now engulfing the world. It commands our whole attention, absorbs our very beings. We face a hostile ideology-global in scope, atheistic in character, ruthless in purpose, and insidious in method. Unhappily the danger it poses promises to be of indefinite duration. To meet it successfully, there is called for, not so much the emotional and transitory sacrifices of crisis, but rather those which enable us to carry forward steadily, surely, and without complaint the burdens of a prolonged and complex struggle-with liberty at stake. Only thus shall we remain, despite every provocation, on our charted course toward permanent peace and human betterment.

Crises there will continue to be. In meeting them, whether foreign or domestic, great or small, there is a recurring temptation to feel that some spectacular and costly action could become the miraculous solution to all current difficulties. A huge increase in newer elements of our defense; development of unrealistic programs to cure every ill in agriculture; a dramatic expansion in basic and applied research-these and many other possibilities, each possibly promising in itself, may be suggested as the only way to the road we which to travel.

But each proposal must be weighed in the light of a broader consideration: the need to maintain balance in and among national programs-balance between the private and the public economy, balance between cost and hoped for advantage-balance between the clearly necessary and the comfortably desirable; balance between our essential requirements as a nation and the duties imposed by the nation upon the individual; balance between action of the moment and the national welfare of the future. Good judgment seeks balance and progress; lack of it eventually finds imbalance and frustration.

The record of many decades stands as proof that our people and their government have, in the main, understood these truths and have responded to them well, in the face of stress and threat. But threats, new in kind or degree, constantly arise. I mention two only.

A vital element in keeping the peace is our military establishment. Our arms must be mighty, ready for instant action, so that no potential aggressor may be tempted to risk his own destruction.

Our military organization today bears little relation to that known by any of my predecessors in peace time, or indeed by the fighting men of World War II or Korea.

Until the latest of our world conflicts, the United States had no armaments industry. American makers of plowshares could, with time and as required, make swords as well. But now we can no longer risk emergency improvisation of national defense; we have been compelled to create a permanent armaments industry of vast proportions. Added to this, three and a half million men and women are directly engaged in the defense establishment. We annually spend on military security more than the net income of all United State corporations.

This conjunction of an immense military establishment and a large arms industry is new in the American experience. The total influence-economic, political, even spiritual-is felt in every city, every state house, every office of the Federal government. We recognize the imperative need for this development. Yet we must not fail to comprehend its grave implications. Our toil, resources, and livelihood are all involved; so is the very structure of our society.

In the councils of government, we must guard against the acquisition of unwarranted influence, whether sought or unsought, by the military-industrial complex. The potential for the disastrous rise of misplaced power exists and will persist.

We must never let the weight of this combination endanger our liberties or democratic processes. We should take nothing for granted only an alert and knowledgeable citizenry can compel the proper meshing of huge industrial and military machinery of defense with our peaceful methods and goals, so that security and liberty may prosper together.

Akin to, and largely responsible for the sweeping changes in our industrial-military posture, has been the technological revolution during recent decades.

In this revolution, research has become central; it also becomes more formalized, complex, and costly. A steadily increasing share is conducted for, by, or at the direction of, the Federal government.

Today, the solitary inventor, tinkering in his shop, has been over shadowed by task forces of scientists in laboratories and testing fields. In the same fashion, the free university, historically the fountainhead of free ideas and scientific discovery, has experienced a revolution in the conduct of research. Partly because of the huge costs involved, a government contract becomes virtually a substitute for intellectual curiosity. For every old blackboard there are now hundreds of new electronic computers.

The prospect of domination of the nation's scholars by Federal employment, project allocations, and the power of money is ever present and is gravely to be regarded.

Yet, in holding scientific research and discovery in respect, as we should, we must also be alert to the equal and opposite danger that public policy could itself become the captive of a scientific-technological elite.

It is the task of statesmanship to mold, to balance, and to integrate these and other forces, new and old, within the principles of our democratic system-ever aiming toward the supreme goals of our free society.

Another factor in maintaining balance involves the element of time. As we peer into society's future, we-you and I, and our government-must avoid the impulse to live only for today, plundering, for our own ease and convenience, the precious resources of tomorrow. We cannot mortgage the material assets of our grandchildren without risking the loss also of their political and spiritual heritage. We want democracy to survive for all generations to come, not to become the insolvent phantom of tomorrow.

Down the long lane of the history yet to be written America knows that this world of ours, ever growing smaller, must avoid becoming a community of dreadful fear and hate, and be, instead, a proud confederation of mutual trust and respect.

Such a confederation must be one of equals. The weakest must come to the conference table with the same confidence as do we, protected as we are by our moral, economic, and military strength. That table, though scarred by many past frustrations, cannot be abandoned for the certain agony of the battlefield.

Disarmament, with mutual honor and confidence, is a continuing imperative. Together we must learn how to compose difference, not with arms, but with intellect and decent purpose. Because this need is so sharp and apparent I confess that I lay down my official responsibilities in this field with a definite sense of disappointment. As one who has witnessed the horror and the lingering sadness of war-as one who knows that another war could utterly destroy this civilization which has been so slowly and painfully built over thousands of years-I wish I could say tonight that a lasting peace is in sight.

Happily, I can say that war has been avoided. Steady progress toward our ultimate goal has been made. But, so much remains to be done. As a private citizen, I shall never cease to do what little I can to help the world advance along that road.

So—in this my last good night to you as your President—I thank you for the many opportunities you have given me for public service in war and peace. I trust that in that service you find somethings worthy; as for the rest of it, I know you will find ways to improve performance in the future.

You and I—my fellow citizens—need to be strong in our faith that all nations, under God, will reach the goal of peace with justice. May we be ever unswerving in devotion to principle, confident but humble with power, diligent in pursuit of the Nation's great goals.

To all the peoples of the world, I once more give expression to America's prayerful and continuing inspiration:

We pray that peoples of all faiths, all races, all nations, may have their great human needs satisfied; that those now denied opportunity shall come to enjoy it to the full; that all who yearn for freedom may experience its spiritual blessings; that those who have freedom will understand, also, its heavy responsibilities; that all who are insensitive to the needs of others will learn charity; that the scourges of poverty, disease and ignorance will be made to disappear from the earth, and that, in the goodness of time, all peoples will come to live together in a peace guaranteed by the binding force of mutual respect and love."

All that we have experienced is an attempt to induce fear for control by the corporate Communist psychopath and the owners of the Military Industrial Complex who direct all governments and democide humans with pollution, harmful cellular / microwave frequencies, and inoculation to provide the pharmaceutical industry a bailout.

I knew the pollution in 2020 would make many people sick and kill them, so did the government and their corporate handlers. We could have really understood the amount of pollution we suffer that year, but they stole the lesson from us by shutting down our daily lives. People were realizing what we suffer from and were waiting to experience the truth, but the government and their corporate handlers wouldn't allow that.

The by-product is the implementation and death of communism and an attempt by the communist handlers (End - Times Zionists) to implement in full swing their superiority and control. To be accurate, the death of the dark controlling force that wishes to control free willed humans.

This same plan of action by the aggressors throughout time on free human beings has been implemented over and over for the theft of the planet, the implementation of the destruction of Mother

Earth and it's inhabitants, to control the consciousness of free human beings, to convince them to destroy their home, each other, and then themselves.

It is true that destruction brings creation, and creation brings destruction. The malfunction occurs in the time frame and their own delusion of being superior and qualified to lead. These energies are impatient because of fear and greed, and have caused the said energies to make many mistakes.

They lack the strength to endure because they are fragile. These energies are less than and lack superiority because of their inability to allow a higher source direct them. Their fear has caused them to frantically grasp the helm not knowing what direction is accurate. This is what an archer refers to as Sin, by being off the mark.

These dark powers are greatly afraid. We need to free them of their fear by removing the control they have. It's too much for them to bear, and their inadequacy is embarrassing.

It is a feud of unity against duality. Yeah there is night and day—it don't take a fuckin' genius to view a cycle. Everything you have ever known and understood is all the one thing. All is not many separate things, but the oneness that subsumes all the parts.

If there is a "mark of a Beast," then that Beast is all carbon-12 reality. It takes 6 protons, 6 electrons, and 6 neutrons to make up carbon-12. Non-organic material (carbon - 13) such as salt and water were already here when the face of God came across the troubled waters.

The word of God is life, and life calmed the troubled waters. We come from Gaia (Mother Earth) and the All Father—we are born of them. The unmoved mover moving in all that moves made us from earth, we are the Earth and the All Father.

Every living being is from the Earth, if there are beings not from the Earth they are trespassers and must explain themselves openly or be put to death. Some humans believe they are from star dust or some bullshit and everyone else is Goy (from clay) indicating that they are from a more pure source. It is obvious to everyone but them that they over compensate for inadequacy. Even God disapproves of their inadequacy and sent a remedy (Jesus) to cure the disease.

That which wishes to democide us with attrition is from the darkness before the light. This darkness needs fear as fuel, and wishes to suck the life from all that is because of inadequacy. It desperately wants to stay alive through theft because of it's lacking in sustenance. In this reality, things that can't live on their own cease to exist.

Ancient writings say that you speak in words without voice to those who dwell down below, and that man is complex and being of earth and fire. Wisdom is hidden in the darkness, and we must be the Sun of the Light without form, lit by the flame of our souls. We must find the wisdom and be LIGHT-

BORN. It's in the heart of the flame, and we must tear open the darkness and shine a LIGHT on the WAY.

The children of shadows are here because man called them to gain great power. There are men who believe in darkness and use dark magic, calling beings from the deep. They are formless of another vibration, existing unseen by the children of Earth-men.

Only through blood could they find form, only through man could they live in the world. A long time ago Masters drove them back from where they came. But some remained hidden in spaces and planes unknown to man, living as shadows but at times appearing to man.

When blood was offered they came to dwell among men. Only by sight did they appear to be men, and are serpent headed when the glamour was lifted. They appear to be men among men and crept into councils as chief priests, scribes, Pharisees, and Sanhedrins. Their arts of slaying rulers and taking the form of rulers over men has allowed them to destroy man and rule in his place. Only by magic and sound could their faces be discovered.

The Masters came to teach man the secret Word that only man can pronounce. The Word is LIFE and in the name of Yeshua all things will come.

These serpents live still resembling man, and walk among places where rites have been said. They can be controlled and bound while in the flesh. If you seek the shadow, evil will appear. Only a master of brightness will conquer the shadow of fear. Fear is the great obstacle, but by being a master of brightness will you make the shadow disappear.

Attempting to reach the Barriers of consciousness is dangerous. But the HOUNDS of the Barrier can only move through angles, they're not free from curved demotions. Only a circle will give you protection from the claws of the DWELLERS IN ANGLES. It is only through strange angles unknown to men do they devour man's soul. Men who dare to reach the Barrier may be held in bondage by these hounds beyond time, held until the cycle is complete when the consciousness leaves.

Make your body into a circle and lose the pursuers in the circle of time. Create the circle that knows not angles, create the form that from the form was formed. Be cautious not to move through angles, for the HOUNDS of the Barrier moves through angles and never through curves of space.

If you leave your body, seek ye to move not in angles but curves. If you hear the bay of the hounds ringing clear and bell like in your being, flee back to your body in circles and penetrate not until you do so.

When you enter your body, use the cross and circle combined, use your voice and utter the word and in the name shall you be free. The circle and cross reminds me of the Medicine Wheel. Seek not to break open the gate beyond, seek your own Light and make yourself ready to pass on the way. Seek to find the light on the way.

This ringing sound these ancient texts refer to has plagued men for a long time. Martin Luther went beyond the barrier of religion and suffered the ringing until his death. Thomas Paine far passed the barrier of monarchy and suffered the ringing until his death. Many of the signers of Texas' Declaration of Independence suffered the ringing until their death. And now hundreds of thousands of people suffer the ringing noise today.

Is it because we have surpassed the barrier of control? What beings have the capacity to emit this ringing noise and convince us to use this technology in our everyday life? How old are these beings, and how long have they been in violation of Universal Law? If these beings know our secrets then why can't we know theirs? Why do they want to destroy us and consume our planet? Because of inadequacy.

If they are alien, did they ask permission to come aboard our existence? If they did, they didn't ask the collective, therefore they are trespassers and have broken Universal Law punishable by death. Present yourself before discovery, it will save you a lot of pain. Is this the reason why our God given rights are unalienable?

Our souls hope for eternal life after leaving these meat vehicles. Some can't believe this and see it as an empty story to be laughed at because their *"possessions of this life are too pleasant, and such pleasures grip the soul by the throat, holding us down to earth. Our possessions posses us."* It is indeed ignorance of the soul.

These possessions in the form of entertainment and communications technologies powered by the dangerous 5G network cause the ringing noise, brain disfunction, cellular degeneration that destroys our very being, deterioration of consciousness, motor function, and cognitive ability. Everyone's want to be the oneness (social media) that subsumes all the parts, and believe this is the only way to achieve the oneness that is desired. This damaging technology has killed and disrupts living beings even before the first broadcast of AM radio.

The new broadcast (5G) is a laser beam with no more wavelength and the same compression and power as a speeding bullet—minus the tangible matter of lead. All "Smart" technology that sends and receives not only harmful Radio frequency, but that sends and receives brain function transmission is immediately dangerous and harmful to life and health of all living beings, and is used to detect all of your wants and desires, then by targeting people's self-image, turning wants into needs.

These devices are also invading your right to privacy, and is used as leverage against you in the court of law, and corporate greed as you voluntary and involuntarily divulge personal information.

All humans became sick instantaneously throughout the world because of a massive amount of air pollution with high concentration that causes the "super spreaders", were large amounts of people became simultaneously sick—falsely thinking it was because of a virus or when they came in contact with each other. Poor water quality in your cells are microwaved and cooked in your internal body, and a dramatic leap in the electrification of our planet with the start-up of 5G are major components of our sickness and death.

Electrification ousted or ionized our sickly electrons, and inflamed our lungs with electrical shock with the conductive high amount of heavy metallic soot everyone in the world involuntarily breaths, and the many vaccines with the severe amount of conductive heavy metals that make humans an antenna to send and receive information while destroying the neural pathways in humans brains cause brain function to diminish.

Everyone involved knows it would hurt and kill people and are a part of an attempted democide that failed because of bad intel. They must all be arrested for crimes against humanity.

The pandemic is a lie, with the help of most all types of media that are the enemy of all living beings, to direct your consciousness elsewhere.

All humans were starting to understand the massive amount of pollution we suffered, most all knew the Kingdom of Saudi and counter parts are the enemy of all living beings. People started understanding there was going to be a timely manufactured economic crisis just as there always is. We saw Hong Kong defeat Communist China and the whole world was on the brink of understanding all the lies we were always told.

Just like 9/11, the dark controlling forces exploited your fears and realigned your consciousness with your slavery, because fear is the greatest obstacle.

It was time for the next manufactured economical crisis, and with the destruction of the opioid cartel, the pharmaceutical companies needed a bailout in form of vaccines and inconclusive false positive test directed by the mass murderers (Fauci, Gallo, and others) to once again (like the AIDS pandemic) draw funds to the Center for Disease Control and many other research groups, with poisonous drugs and large amounts of grant money for research.

It also helped our exit from Afghanistan for no longer needing to guard opium fields from which we obtained the opioid.

The Federal Reserve was always going to dump the market and I've been saying it for about two years prior. The Federal Reserve and its fractural lending is fraud, theft, and controls the market at its whims. The Federal Reserve can manufacture any market that suits them.

When oil is spewing into the ocean, killing ocean life, the price goes down because of perception of emotions. If they claim the pipeline has been attacked the price goes up. Refineries are running at 100% capacity at all time. There is the same amount of fuel at all times, the only thing that will change the price is your perception, emotional fear to hoard, and what price they can get out of you.

These multi-directional attacks are to democide humans because of overpopulation, halt the consumption of materials necessary to sustain life, to bring massive amounts of money to Satanic and greedy corporations, factions, shadow governments, and kingdoms undeserving to survive.

With the help of many people like Edward Bernays, we have been bred to consume and waste the rest, and get back on the treadmill for more slavery. Most all humans have been bred to destroy their surroundings, each other, and then themselves.

Gun sales are up, and one of the things keeping us from complete corporate slavery are a bunch of armed rightwing nut jobs. Corporate psychopaths only take what they know they can get because they are cowards. Even liberal Californians purchased a massive amount of guns during the pandemic—smart.

It was mainly the faithless that had the most trouble and were easily controlled.

I am ashamed of how most all people reacted, including all the tough guy patriots. Crimes against liberty, freedoms, and humanity have been committed and you did nothing.

I asked people to at least talk about punishing all violators of the Constitution of The United States, and most were too scared to even consider it quietly in their own minds. I ask you to talk about what we can stand for and what we can not, and you did nothing.

I asked for you to consider a plan if the aggressor physically took action, and you did nothing. I explained to you that Sam Houston defeated the Mexican army with less than 900 pissed off Texians, most were elderly and more hardy than you panty wastes.

I explained to you that we have the right and <u>duty</u> under the Constitution of The United States to form an organized militia to combat enemies foreign and domestic, and you did nothing.

How do you ever expect to become more free if you're not willing to stand up against tyranny in any form and lay your life down if necessary?

Dying for freedom doesn't mean you are willing to go to another country and kill people who you view as aggressors on freedom—it means you would rather die than be unfree.

We had misinformation from most all media sources in an attempt to overthrow a political faction, causing great mental and emotional anguish to all viewers. All the while claiming misinformation from anyone who didn't tag along with the narrative.

An attempted compulsory vaccination forced by government, peer pressure, and social media—corporations that prohibited the ability to purchase food freely and enter the public accessible facilities, and the forced vaccinations to keep one's job.

Mask enforcers such as police officers in dereliction of their constitutional oath arresting and beating up old ladies and little girls, Communist indoctrinated governors producing edicts repugnant to the constitution, judges throwing people in jail for attempting to keep their business afloat, and the destruction of most all small business—and you stood by with your hands in your pockets.

"All laws which are repugnant to the Constitution are null and void." Marbury vs. Madison, 5 US (2 Cranch) 137, 174, 176, (1803)

"Where rights secured by the Constitution are involved, there can be no rule making or legislation which would abrogate them." Miranda vs. Arizona, 384 US 436 p. 491

Which means a great number of police officers, judges, and governors are guilty of crimes against the American people and must be punished.

We should have already brought down great pain, violence, and furious anger upon all said entities that were in violation of the Constitution of the United States.

The only governor who made a real leap towards freedom is DeSantis of Florida, who signed a bill banning vaccine Passports, and canceled local covid rules.

Democrats want every human to show proof of vaccination, but are against showing your identification to vote. Their handlers care not if these people get lynched or reelected, just so long as they do what they are told. We have a plethora of brain dead politician want-to-be's that would suck a dick on television to become famous.

Vaccination passports are what the Gestapo ask for "papers please." These very people who hate Adolf Hitler and are in love with World War II are the very people who imitate the NAZIs, ironic.

The Texas governor, in vain, wanted to look like a real patriot after the fact, but most all other states and officials are in violation of the Constitution of The United States and must be brought to justice. They can either have their liberties arrested or we must physically punish them ourselves by whipping them accordingly. There must be a center to bring evidence of violations so we can root out these violators—governments, businesses, or individuals.

This was a litmus test to see where we stood in freedom. It doesn't look good.

The Declaration of Independence says it's our right and <u>duty</u> to abolish any government that doesn't secure our unalienable rights, including life, liberty, property, and the pursuit of happiness. How the fuck do you reckon you abolish a compulsory armed and trained security force? With force and weaponry.

Palaces are built upon the bowers of paradise.

The government has a valve on normal life and can turn off and on the privilege to do so depending on your obedience. The U.S. Constitution is the supreme law of the land, and any statues, to be valid, must be in agreement—we do not agree. It's <u>impossi-</u>

ble for a law which violates the Constitution to be valid. You must declare your God given rights from birth.

This is succinctly stated as followed:

I GIVE NOT CONSENT TO ANY VIOLATION OF MY GOD - GIVEN RIGHTS AND FREEDOMS. I GIVE NOT POWER TO VIOLATORS, OPPRESSORS, TYRANTS, AND EVIL ENTITIES. I AM PROTECTED BY MY CREATOR, AND DIVINE FORCES.

The only purpose for government is to protect personal right. The hierarchy is : 1) God, 2) humans, 3) government, 4) corporations and the list goes on. You are not a citizen, resident, straw man in bold capital lettering on your social security card and birth certificate. Do not respond to these labels. You need an affidavit of statute and they have to try to rebut it. They can't and have no standing in the matter.

A county, city, or state is not an individual nor can a corporate fiction have power over you. Mr. and Mrs. State of Texas doesn't exist. You have the right to face and challenge your accusers.

These aggressors have not jurisdiction over you. I know it's hard for you to understand because we live in a world of opposites. We have a natural right to travel and policies are for people that work for the corporation—employees only. It's not a law and can't be implemented. These compulsive rule followers hold the state of slavery in place.

The government wants you to wear a mask not only for compliance, but to continue sulfur dioxide metallic soot release. Wearing a cloth mask will help you survive longer in the industrial soot, but will not correct the root cause—sulfur dioxide and ground level background ozone.

Most of you people can't distinguish the difference between government and society. Our society is based on our wants, and government on our wickedness. One unites our happiness and affections, the other restrains our vices. People subconsciously want the USA to fall, when it's evil that they really want to eliminate.

A general <u>misconception</u> is—any statute passed by legislators bearing the appearance of law constitutes the law of the land. Mandates of any kind is just an appearance of a law exploiting your ignorance.

Where it be a passive ruler or a tyrant means nothing to me. Life, liberty, and the pursuit of happiness, and owning your person and property (one and the same) supersedes this gutter filth.

We must mandate and legalize freedom. Democide isn't necessary. We can control our own population growth without being murdered, poisoned, or psychologically conditioned to be self-saboteurs. This is where personal responsibility gives people the opportunity to make good and bad decisions.

If someone doesn't want to wear a seatbelt, then don't. Seatbelts save lives, some people's lives aren't worth saving. Will they lose control and hit someone else? Maybe. It's your own responsibility to drive defensively and know that tomorrow isn't promised. Maybe no one should drive anymore, so we can stop car wrecks.

Anyone who gives up their freedom for safety deserves neither.

If you don't want to wear a helmet while riding a motorcycle, then don't. Apparently there are too many people on the road anyway. You have the right given to you by God to make good and bad decisions, it's called free-will. You have the right not to drive on the road as well.

If you want to wear a mask and get vaccinated by an experimental, unapproved vaccine then you have the right to have injected anything you want, including harmful drugs or poison. You have not the right to force anyone else to do as you do.

My great aunt wore a mask all my life, poor thing thought her allergies were causing her trouble, when it was actually a massive amount of pollution that caused her breathing problems. I was already accustomed to it and thought of her.

And now I look at you people alone in your car wearing a mask—swimming under water wearing a masks. We want people wearing mask, so we can easily identify people who can't think for themselves.

Like the cult of Qanon, you wear the mask in solidarity against former President Trump, you are in a cult. Your desire to be perceived as a good person

has out weighed your intelligence. You lack the ability
to think for yourselves in fear of being wrong. So you
do what everyone else is doing because you don't
know any truths. I indeed have pity on you because
the number of lies far surpasses the number of truths.
The truths change not because they are Universally
true through out time; classic you might say.

XXVII

Let's talk a little bit about Destructive Religious Cult.

A Destructive Religious Cult or DRC is any group which uses psychological manipulation to impair, destroy or make captive an individual's freedom of thought or reasoning abilities. This is done with the hidden purpose of promoting the wealth, power or vanity of charismatic cult leaders.

Sounds like Facebook, most all media groups, all political parties, wokes, Qanon, government, school of all levels, LGBTQ+, a lot of religious organizations, shit... our entire life is a cult designed for control.

The recruitment practice is usually deceptive and the victim enters the cult without informed consent. Cult leaders demand blind faith in their teachings, restrict the freedom of their followers, and direct them to engage in criminal activity. Sounds familiar, Communism is a powerful tool of control.

Cults use techniques of psychological manipulation of isolation. There's a loss of reality induced by physical separation from society and rational reference. Stay home and social distance if you care about lives and are an upstanding citizen.

Facebook can give a prosthetic feeling of togetherness while being separate and has cults within cults. Each subscriber has followers with approvals and unwanted input or worship.

The use of and being in the state of high suggestibility induced by hypnosis, often thinly disguised as meditation. Yes, the media uses suggestion while luring us to sleep.

Peer Group Pressure used in suppression of doubt and resistance to new ideas achieved by exploiting the natural need to belong. Sounds like the vaccination hesitancy policy Facebook implements, along with most media groups, government, and everyone who was in fear of going against CDC guidelines.

Removal of privacy and the loss of ability to evaluate, logically achieved by preventing private contemplation. Yes, Facebook and most all technologies invade our privacy. Your cell phone can detect what you're thinking and counteract social discourse with social engineering.

Games and the need for direction when playing games with obscure rules increases dependence on the group. Sounds like company retreats.

Meta communication with subliminal messages implanted by stressing certain key words or phrases in long confusing lectures. Vaccination is Vaxx a Nation. Also, Facebook is rebranding as Meta, if that gives you any indication.

No questions with automatic acceptance of beliefs accomplished by discouraging questions. Sounds like the government and the CDC rules and guidelines. Scientist have all the answers and know more than us, no need to question them, why else would they wear a white coat?

Confusing doctrines with complex lectures on an incomprehensible doctrine, encourage rejection of logic and blind acceptance. Sounds like Fauci's interviews, and most "scientists" delusions.

Rejection of old values and the acceptance of new life style accelerated by constantly denouncing former values and beliefs. Must be the new normal everyone is referring too, or the "Great Reset."

Confessions with the destruction of personal egos, increased vulnerability to new teachings and recruits' weaknesses revealed, through sharing innermost secrets. Like your post on Facebook and Instagram, also communists are not individuals—they are a part of the hive or cell.

Flaunting hierarchy by acceptance of cult authority produced by promising advancement, power and salvation. You can get to many levels of Qanon depending on your acceptance. Also, corporations tell you to hang in there. We're going to get you that raise.

Finger pointing with a false sense of righteousness created by pointing to the shortcomings of the outside world and other cults. You didn't get a vaccine? You are the down fall of our society. Also, Psychological Projection of most all media groups.

Individuality removed by demanding conformity to the group dress code. Wear your mask or the colors of the rainbow.

There are 27 different steps in the recruitment process. First the indoctrination through schools and disorientation process with cognitive dissonance.

After the Recruitment and Indoctrination phases, physical and mental changes appear in the Disorientation state. Of the total 27 steps of psychological coercion, only 17 need to be used in a 48-hour period to result in a disoriented person.

People who think for themselves are usually singled out, while idealistic people whom are accepting of their new friends' ideas are put on the fast track.

It starts with a mental pause to a snapped person. Then two categories develop: 1) dysfunctional brain response, 2) acceptance without wisdom or judgement.

Dysfunctional brain response consists of confusion, change in physical appearance, emotional distress, no self reflection, change in speech pattern, and information process destroyed. This is when there might be some important truths, but you're not sure.

Acceptance without wisdom or judgment consists of personal goals, imagination equals reality, fact is equal to fantasy, accepting leader's thought, and conformity. Qanon continues to direct people with what they think should happen with our government, but is not. You must learn and adopt what Xi Jinping says, learn the People's Republic of China, believe.

One side is high excitement which consist of low food intake, love bombing, noise like ringing sounds, continued monitoring, and lack of sleep. Thousands of people hear the ringing noise and can't sleep, our devices monitor our every movement, habits, and thoughts.

The other side is information distortion which consists of meditation, guided fantasy, subliminal oratory, trance lectures, mis-directed thought input. The misinformation of media groups directing your consciousness with distortion, many variety's and flavors of the lie. Demons move in angles and can circle not, they have nothing to circle back too.

Then recruitment with ego destruction which consists of guilt, fear, and threats. And Identification destruction which consist of isolation, personal conformation, restricted sex drive. Sounds like most Millennials.

Brainwashing: Often in conjunction with drugs, is sophisticated hypnosis which involves the associative pairing of induced pain/terror, the cult message, and trigger cues. Trigger cues are planted in the subconscious and are too numerous to list.

The government induces pain/terror through media and social media telling you over and over to come together and stand against terrorist groups or viruses as control over your consciousness. They use trigger cues like 9/11, Bin Ladin, CDC, Fauci, and appear to be used from separate angles from political parties. These cults are one and the same, controlled opposition to redirect your thoughts.

For instance, flags and hats that say "Make America Great Again" are triggers for people who worship Former President Trump and for people who despise Trump. Most all are multifaceted, and most all are controlled by the cult that is our government. The main reason for cults is control.

The triggers are later utilized by the cult to control the survivor without his or her conscious awareness using visual symbols on greeting cards, flower colors and arrangements, common hand gestures or the covering of one eye, verbal phrases, body postures, facial movements, various types of flags and symbols, and the list goes on. The survivors can be placed in high ranks of authority and be controlled with ease, or fear of exposure.

Just as trigger words can coerce you to act or not react, the usage of shut-up words can also shut people's reception down. Words like racists, critical race theory, Qanon, conspiracy, conspiracy theorist, supremacy, terrorist, hate crime, homophobic, Trump, woke, conservative, liberal.

All words that give completion in thought while preventing any more information obtained. You already have an idea pertaining to the word and no further explanation is required, or they're trying to convince me to switch teams.

For instance, telling people Qanon is a cult is ineffective because people who are against former President Trump instantly think you're trying to get them to join, while people who are in Qanon think you're the enemy for obvious reasons. Simply saying Qanon shuts brain function down completely, or turns on the parts of the brain that have been infiltrated.

Whether Qanon tells some truths or none is meaningless to the control it possesses over one's consciousness. Brainwashing is an integral part of ritual abuse and cult indoctrination, and also serves to create amnesia for cult information such as names and places to protect cult secrecy. Governments perform this as well, they can't have everyone knowing they are bullshit and stealing tax dollars.

Hypnotic introduction of visual images during abuse can hamper later therapeutic efforts to uncover accurate memories. These tactics are thousands of years old.

There are so many variations of the US flag that the actual flag is meaningless. The American flag is distorted to look like the gay flag, the police flag, the backwards flag (not meaning you're going into battle, but the opposite of what America is founded on), the zombie shredded flag, and numerous other flags.

Worshipping a flag is as good as worshipping a golden calf, it's now used as an idol to stand for or against. You might as well worship your driver's license because that's what it is, identification and nothing more. Flag code must be enforced for true identification.

Cult indoctrinators teach cognitive confusion as the oppositions of what the inductee views as good or bad are constantly reconciled in cult behavior and training, such as, be obedient/rebel, sex without pain is not pleasure, love and hate are the same, black is white, ugly is pretty, left is right, living in an alternate computer reality (Meta) is better than reality itself, monsters are angels.

We have been living in a world of opposites. It causes children to split inside to function within this cult confusion and the non-cult world with different rules at home and social gatherings. It is a feud of unity against duality.

The purpose for cults is to control with the hidden purpose of promoting the wealth, power or vanity of charismatic cult leaders.

The want to control the meek and mild has been desired from the first time a human wanted to preserve a child's life, later to manipulate the child to preserve their own life when becoming elderly. When you're very young or old it's a great feat to exist on your own and requires help from willing and unwilling participants.

The unwilling need conditioning established early because in later years the elderly have not the strength and mental capacity to force the unwilling participant. Examples through your own actions are how we voluntarily condition all participants. Examples of how to manipulate are also inherited and improved.

The word cult stems from the word culture (intellectual achievements, customs, or in conditions suitable to grow) and later took the meaning of worship.

Most cultures have an ideology developed early in their evolution. Within that culture are subcultures that take the prescribed or subscribed thought and angles it to the direction of control with the hidden purpose of promoting the wealth, power or vanity of charismatic cult leaders.

The cult of Mithras is a cult for the military that has a complex system of seven grades of initiation and communal ritual meals. Initiates called themselves Syndexioi, those "united by handshake." You swear an oath to secrecy, dedication, and recital of their catechism. Their temples are windowless and underground, much like the Masonic Lodges of today.

While Moses was on Mount Sinai receiving the word, waiting at the bottom—chief priest and Sanhedrins were making their own laws like the Talmud. It is claimed much later by Rabbis that Moses was given two Torahs, one written one oral. The oral were laws that justified evil men's actions and not for everyday viewing so as not to incriminate.

For a long time the Talmud was spoken word only because if Moses had known of this Talmud he would have done more that smite them and destroy their precious golden calf. He would have killed them for laws and acceptance like the raping of animals or children of ages six months and older. Or the idea that it's ok to steal, swindle, and leave Gentiles at the bottom of a well.

Hundred of years later (around 200 CE), a Rabbi named Judah I was the chief editor and redactor of the Mishnah (written form of the oral Talmud), claimed it was an oral Torah given to Moses by God and was passed down to Joshua, Judges, more reliable sources not explained, finally to the Pharisees who caused irreparable damage to society for control of one's everyday life, not law. Eventually the Jews extract from scripture, other times they extract from the Talmud. It will be Jewish law or you're pushed away from the security of the community.

The laws were fluid as the Rabbis chose and decrees induced more restriction, but only God can make laws. Rabbis could strengthen or relax laws as consensus desired, depending on the gossip of the time.

They claimed this oral Torah would answer questions of everyday life. Rabbis in Galilee and Babylon wrote commentary in the Talmud, like a continuous discussion. There was a Palestinian or Jerusalem Talmud and a Babylonian Talmud. The Babylonian Talmud is the controlling source.

One of Jacob's descendents and tribe was favored in Egypt, then in a later shift in authority caused them to be enslaved. This left a bad taste in their mouths from their conquerors and most took up paganism. This learned indoctrination of the Talmud continues to this day.

Christians, Jews, and Muslims are all family members and have the same father, Abraham. They all worship the same God as well.

It was the covenant of Abraham which developed these religions. Abraham came from a family that worshipped many gods and performed human sacrifice, but this time he will worship one god (Yahweh) and be circumcised. These two things are the second covenant. The promise of his seed to be as numbered as the stars in the heavens if he worshipped one.

Worship meaning, using external signs and acknowledging it through sacrifice. Usually through killing and burning so the sacrifice will be emitted into the air. In the ancient world internal contemplation wasn't enough. Worshipping means who you thought had power and what you did to acknowledge it.

After the beginning, the tower of Babel, came the flood (the first Covenant with Noah), the story continues with attention on Abraham and his descendents.

Abraham, Sarah, and the tribe of nomadic sheep herders wandered through Canaan, Syria, and Turkey following the sheep herd. It was then that God spoke to Abraham and promised if they kept the Covenant his numbers would be many and they would have a land of their own from the river of Egypt to the "Great River," the river Euphrates.

We depend on numbers for survival even to-day—God is promising the survival of his family. In no way would nomads find the idea of farming glorious, quite the opposite. The word Hebrew means "to wander."

Abraham and Sarah give laughter to the idea of having children at a very old age. Sarah proposes Abraham have a child with his Egyptian slave girl Hagar, and does. His first born son is Ishmael, like God promised.

Later, Sarah's jealousy sets in and she doesn't like the idea of this Egyptian slave girl and her bastard child in the camp. Sarah tells Abraham God told her they got to go. They are sent off to their deaths in the desert of the Negev, which is Southern Israel.

God promised that Ishmael will be the father of a great nation, and is indeed the father of the Arab nation. The Book of Jubilees explains that not only Ishmael, but his cousins (from another wife of Abraham, Keturah) also became the Great Nation.

Ishmael and his mother almost die when a spring of water mysteriously comes forth called today the Zamzam Well. This is the end of Ishmael's character in the Torah, which is the first five books in the old Testament or the Jewish scriptures, but not in the Quran or the Book of Jubilees.

The Book of Jubilees explains when Abraham was on his death bed, he summoned all of his children including Ishmael and offspring. Abraham wanted everyone to continue to observe the traditions and marry not the Canaanites. Abraham sent Ishmael and offspring to settle "between Pharan and the border of Babylon, in all the land to the East, facing the desert. And these mingled with each other, and they were called Arabs and Ishmaelites" (Jubilees 20:11-13).

Muhammad, the prophet of the Muslims, says Abraham and Ishmael made the Kaaba in Mecca. Muslim tradition says Abraham began to miss Ishmael and Hagar and searched for them. The sacred spring is in Mecca, Hagar running around looking for water is the ritual pilgrimage where pilgrims run back and forth between two points in Mecca and derives from the story of Ishmael and Hagar.

Abraham's son Ishmael does have a land of his own from the river of Egypt to the "Great River," the river Euphrates, while the others wander.

Around the second or third century CE, the Christians and Jews understood that the Arabs are the descendants of Abraham. They had the same traditions as the Jews and Christians and were known as the Ishmaelites. Jews and Christians were at the time the same people, it was the conversion of Jews to Christianity which took power away from centralized authority. Later the new inheritors, the Gentiles through Christianity, arrive.

Ishmael has nothing to do with being Muslim, following the preachment of the prophet Mohammad from the Dissertation is where Muslims arrive. Being Arab is an ethnic identification, being Arab means not that they are Muslim.

Abraham returns to the biblical story when his second son Isaac is born. The story continues through Isaac's linage.

God tells Abraham to sacrifice Isaac, it is said to test his obedience. At the last minute God intervenes and shows Abraham a ram more suitable to fill in for Isaac. The request was canceled by God himself.

If this was a way to make Issac obedient, it failed miserably and caused ripples of negativity throughout time.

Paul says it was Abrahams faith that saved Isaac not his obedience, it was his faith that Abraham received the Covenant. It looks to me that Abraham was one of the first to say enough with the human sacrifice. Abraham dies and is buried in Hebron south of Jerusalem and Bethlehem, and Isaac has a son named Jacob.

Jacob is the second born as Esau is the first son of Isaac. Under devilish means, Jacob believes he's stealing Esau's birthright by acting like he is Esau, and Isaac gives Jacob his blessing on his death bed. This lie and theft nullifies most all things he does from here on out.

Jacob travels and sleeps at Beth-El (House of God) and sees in a dream the heavens open up with a ladder descending. He sees people traveling back and forth from heaven to earth. He anoints a stone as commemoration.

All this happens in a dream, but how do we know? Jacob is a deceitful liar and thief. He changes his name to Israel and the Children of Abraham are now known as the Children of Israel or Israelites. Jacob even stole his grandfather's honor.

The word "Jew" later derives from where they are from, Judaea. Which would tell most independent thinkers that being Jew is like being a Texan from Texas. Being an Israelite is being a descendant of a liar and thief, and there is no way to convert to being a blood relative; all other humans are Goyim. Goyim meaning from clay or idol worshippers—less than, but at first used as the meaning of nation. I wonder where the Edomites of Esau's are? Most likely with the Greeks.

Jacob has twelve descendants which are known as the twelve tribes of Israel. One of the descendants is favored in Egypt, probably because their cousins (Esau's children) are with the Greeks. Then there was a change of rulers and they were oppressed.

Now enters Moses. He is instructed to tell the Pharaoh to let his people go. It is explained in more detail in the Quran of how Moses negotiated and showed signs of divine relations and miracles. They flee Egypt and the Pharaoh's army drowns on the attack.

Moses makes his trek through Sinai where he is summoned to the top of a mountain and receives the Torah or the first five books of the Bible (Jewish law) and the third Covenant. It is more in-depth than the first two Covenants. This is good because they are in desperate need of clarification. Moses never sees God's face, but after the experience Moses was said to glow.

The Ten Commandments is the overview of the new laws but is broken in a fit of rage from what Moses witnesses. After all he has done, they worship a golden calf. I don't blame him for beating them or breaking the stones. I would have rather him kill the Sanhedrins, this would have eliminated great pain for humans.

Moses climbs back and asks if God can do it again, and does. Moses and all with him wander for forty years and almost die many time but are saved by God from starvation with Manna for food that only grows in the morning and evening, and dehydration when water springs up when Moses taps his rod on a rock. Moses has an expectation that there will be another like him, Jesus and Mohammad.

It is important to know you can walk from Egypt to Lebanon in about three weeks. It's close to six hundred miles, less than walking across Texas (from Orange, Texas to El Paso, Texas is 888 miles), and had more vegetation and water than present day.

There was ample time to find the "promised land," but Moses found it not.

Moses dies at the end of the Biblical five books (the Torah) and they never find their promised land, even though Ishmael does. The tribe moves into Canaan (Israel) where people already live, Ishmael's heirs. They kill the inhabitants and occupy the land like they do today. They claim it was fulfillment, but Moses, who was more in contact with the Lord then they, never settled there. Even after traveling through many times did he not. Moses could have ousted the bad apples at Sinai, but he's only human and could have never known the outcome.

God evolves in the story as well. God is less hands on as time passes.

It is here that the Jews now want a king and are in love with obedience because they can't think for themselves. It is said that God reluctantly allows this, first Saul then David.

David was favored by the lord and kills Goliath, conquering Jerusalem making it the capital and spiritual capital by building a massive temple which he doesn't finish. David is now the monarch of the Israelites. David's lust and pride began to dwindle the Lord's favor. Because he was king he sent a man to his death because he coveted that man's wife. As punishment, David's first born son dies; his pain doesn't last long because he cares more about himself. He can get a new family.

It is Solomon, the second born, that inherits the monarchy and finishes the temple. Solomon was claimed to be wise and had 700 wives and 300 concubines. One of the wives was the Pharaoh's daughter, others were pagans from other places, like Ethiopia, that brought their beliefs and traditions that turned Solomon's heart away from God. It was claimed by the Ethiopians that Solomon had a child with the Queen of Sheba.

Solomon was known as the Patron Saint of Magic, wearing a magic ring with the seal of Solomon which is now referred to as the Star of David. This ring was said to give power over demons and angels that brought him precious stones and built the temple, stonemasons. It was a Medieval symbol of invoking magic, mainly to divide as he choose, the Star of Divide.

Solomon developed ceremonies and sacrificial cults in the temple which also served as the corporation and depository of riches. What was this cult sacrificing at this point? They were rich and powerful enough to sacrifice anything they chose, including humans. This temple was the dominating force in Jerusalem. Paganism crept in from David and Solomon, and an evil cloud came over Israel. All under David and Solomon became as evil as the times were.

As punishment for turning against God, the Tribes of Israel are removed, Solomon dies, and everything unravels. The kingdom splits to Sumerian and Mount Zion (the City of David), the king of Israel to the north and the king of Judaea to the south. The southern kings write the Bible, it's the Judaism account.

The Babylonians arrive and conquer the Israelites and God doesn't intervene. The last time God intervenes is 721 BCE against the Assyrians, and the Israelites are now on their own. The Babylonians loot the temple and most likely the Ark of the Covenant and enslave the Israelites again.

It's only the most skilled craftsmen and upper elite that are carried to exile, the Babylonians have plenty of peasants. It is called the Babylonian Exile, two generations (60-70 years) disbursed. The only place you could worship and sacrifice was in the temple of David and Solomon, this changed the style of worship from physical signs to prayer alone.

The Babylonians are replaced by the Iranians, namely King Cyrus. Now they can return but most Israelites stay. From about 550 BC to the 1950s, the Israelites have a continual community in (Babylon) Iraq.

It was in the 1950s that America founded the new Israel. America was becoming friendly with the Arabs because of oil exploitation. The Arabs wanted the Jews to be removed and America wanted to lend a hand as they did on more occasions. America continues to pay ignorantly to this day because they have been doing it for so long not knowing why America pays Israel ten million dollars a day, it's now just a habit. It can't be because they're kinfolk to Jesus,

God and Jesus turn away from the Israelites, the sons of Jacob.

When they return they have two new institutions, the Synagogue is for prayer and the new professional Scribes "experts of the law," but most move back to Judaea. The study of the law is done in the Synagogue; the Synagogue determines what the Scribe learned.

Ten of the twelve tribes disappeared, everyone that is left is now referred to as Judeans or Jews. It's a broad term that now covers all. The Jews that stay and assimilate are called Samaritans.

The Jews live under Persian rule for about one hundred years. Around 300 BC Alexander the Great arrives and destroys the Iranian Empire. Alexander doesn't live long; upon his death the generals divide up the kingdoms (Egypt, Syrian, Iraqi, etc.).

The Hellenization of the Jews (culture, lifestyle, point of view) threaten their beliefs. Many Jews found the intellectual position of Hellenism attractive.

The Greeks say someone can understand the universe by their own means, the monotheists explain it by God. It's the evolution-creation debate; believe me when I say there is nothing new about humans. The by-product is sectarianism and around 300 BC there is a great divide. This sectarianism birthed the Pharisees, Sadducees, Essenes, Zealots, followers of John the Baptist, and the Jesus movement.

The Pharisees had strict observation of the law or the tradition of the father. These are the people who killed Stephen and many others for words spoken, believed to be blasphemous. Paul was a Pharisee and was granted power by the Synagogue to hunt down Christians but was blinded by a vision of Jesus in Damascus.

The Sadducees were high priests, aristocratic families, and merchants which said if it isn't law you could do as you choose. You're bound by scripture alone. Kinda like communists—the top of the pyramid is capitalist, everyone below has to live communist or under the hand of the Pharisees. The Sadducees were literal interpreters of the law and cared not of the "traditions of the Father."

The Essenes were a mystical sect that possessed no money, were priests, didn't marry, and practiced Asceticism. Asceticism is a lifestyle of self-denial and discipline for religious ends. It is claimed the Dead Sea Scrolls came from the Essenes' library. They rejected the Hasmonean (high priesthood in Jerusalem) control and participated not in the formula of the Temple.

These groups were the Philosophical schools prescribing ideology. They were all trying to redefine themselves like we are today.

The Zealots were revolutionary priests involved in the insurrection of 66-70 CE. Kinda like Oliver North and Admiral Poindexter, operation without guidance on their own behalf.

These groups were more like political parties—Republican, Democrats, Conservative, Liberal, Progressive.

The followers of John the Baptist preached the end, repentance, and laid the ground work for the Jesus movement.

The Jesus movement was later called Christianity and preached the Kingdom of God is at hand. This movement was very different from anything that has been witnessed, it was the first time there was a social uplifting message and the understanding that God is for all including the Gentiles.

The understanding that the sons of man had authority over demons, satan, and the sabbath—meaning even death changed not the powers and authority of the righteous sons of man that was given to him at birth. Jesus fulfilled the Old Testament scriptures of Isaiah's prophecy, and the Jewish priest engineered Jesus' death.

Many accounts besides the four gospels indicate Jesus cast out demons, cured the sick, raised the dead, and multiplied fish and bread to feed five thousand. None before had performed such miracles and remained humble. Most others wanted to become king.

It was Jesus that saved us from senseless killing by the Jews because of the letter of the law. None of the Jews healed the sick, raised the dead, or removed demons from humans, nor cared about anyone but themselves.

One of the many ways Christians and Jews clashed is because Jews wanted Gentiles to become Jewish before becoming Christians, they wanted the middleman power to go through them to get to God. Jesus saved us from Jacob's superiority and diminished their authority. The Jews, Scribes, and Pharisees wanted the letter of the law because they lacked the ability to understand why there are laws.

The intention of laws is to make an understanding of intention, chewing food before you swallow is smart. Making a law that says you must chew your food 20 times before swallowing or be stoned is ignorant, as many people are in this existence.

It was Paul's explication that cleared the way for Martin Luther who understood we didn't need to pay money to the Catholic Church to get into heaven. It will be easier for a camel to pass through the eye of a needle than a rich man to enter the kingdom of God.

The Jewish Diaspora, or the scattering of the Jews, caused them to be exposed to outside influence. They were originally nomadic but kept to themselves in clans.

A new despair arrived and the Romans gave the Jews what they desired, a king. Herod isn't exactly a Jew and comes from the Diaspora with outside influence.

Herod is a go between and keeps order and peace. The Roman oppression caused everyone to desire a final solution or Apocalyptic "unveiling" and hoped for an "anointed one" or messiah. Others wanted to restore the Kingdom of David and reunite the twelve tribes, but God allows not for this to happen. Herod dies and at the same time Jesus is born.

The Jews wanted Rome to acknowledge the one true faith and reunite the twelve tribes to bring about the end times resolution.

The Jesus movement grew with Roman oppression, despair, expectations, hope, and political intervention. Christianity is a Jewish sectarian movement, the Jewish past is the Christian and Muslim past.

Dissent between the Hebrew and Greek speakers arises where it's easy to identify the enemy. This led to the stoning of Stephen, a greek speaking Jew from the Diaspora who converted to Christianity and preached the word.

Saul, a greek speaking Hellenized Jew, witnessed Stephens preachment and stoning and even held participants cloaks. He wants to hunt Christians down by the authority of the synagogue. Saul has a blinding vision and hears words from Jesus and has to be led directly to the Christians; he is still blinded. Later he changes his name to Paul and preaches the word to the Gentiles of Syria without the circumcision or conversion to Judaism.

Jesus is written about is the Gospels, or "the Good News," and is called the messiah even though he is reluctant to receive this. Messiah means the "Anointed One," but Paul writes about Jesus and says "Christ" like it's his last name, Jesus Christ. Jesus is an observant Jew of his time and didn't come to abolish the Torah.

The Maccabees write a book legitimizing their one hundred year kingdom that is written like propaganda and the Romans couldn't tolerate it.

A devotion to Mary arrives saying she is a virgin before and after Jesus, his brothers (James) and sisters are now considered cousins. But when Jesus returns home they all think he's crazy and deny him. They said this is Jesus? Joseph and Marys' son, right? When Jesus dies he tells Mary "behold your son" meaning look at me now.

Religions today still behold to Mary, even after she denied Jesus several times, like when he returned home. Roman influence made the mother important from Romulus and Remus and their she-wolf mother.

The only virgin birth mentioned in the four Gospels was in Luke—he was Greek. In all other religions the virgin birth occurred (Leda and the Swan, Persephone and the serpent).

Mary had Jesus change water into wine at a wedding because they had none. So it wasn't like she didn't know he was more than. The height of this ideology was in 1854 when Pius IX proclaimed the Immaculate Conception "by our own authority." It was defined as dogma by the Roman Catholic Church in 1950. Other than the birth of Jesus, Mary doesn't play an important role in the New Testament but her virginity is affirmed in the Quran.

Now comes the Acts of the Apostles by Luke. Judas Iscariot's (who killed himself for betraying Jesus) position is filled to bring the number of the apostles back to twelve. Their Jewish heritage and tradition's continues, but they are doing it on their own accord, without Jesus.

The twelve congregate in the upper room and they birth the church with a spirit descending upon them. It's Peter that Jesus will build his church upon, but not complete.

In the Diaspora synagogues, the Gentile sympathizers are invited to attend in a marginal way. More strife arises when Paul says the Gentiles need to adhere not the the Torah. Paul argues that Jesus rendered the Torah unnecessary. There was a large influx of Gentiles into the Jewish-Christian community with equal footing. This was a major reason for the separation of Christians and Jews. The Jews didn't want Christians and Gentiles to be inheritors of the spiritual Covenant, the Jews saw it as a take over.

The Jews have a double standard and are inflamed from this. The Jews were exempt from veneration to the Roman ruler or military service, but still think no one is exempt from the Jews.

Word gets back to the elders of strict observation and the followers of Jesus are in Jerusalem. Paul is summoned to explain himself.

Part of the end times scenario is the gathering of the Gentiles that adopt the belief of Jesus. Everyone thought it was the end times, so they waited.

This divided the Jews and Christians. The Jews woke up the next day and it was the same, and their expectation was met not. The Death and resurrection marked the Messianic Spiritual Era, but the Jews were looking for a second coming. The Jesus movement became Christianity, its own religion with the gathering of Gentiles.

The Jews didn't like that the Gentiles were not circumcised nor observed the Torah, and were now the heirs of the covenant. Jews say they ain't Jew at all, the Christians say fine we ain't Jew we're Cristiano. And now the parting occurs.

The Jews are crafty and have learned ways of manipulation from David and Solomon to pour poisonous half truths into ears to control the human mind. These techniques are handed down to the Masons of today.

The Jews convince the Romans that the Christians are not Jew at all, so they are not exempt from veneration of the ruler or military service. Rome was in an economical depression and needed a diversion, so they brought on the circus and fed Christians to the lions.

Like the Jews attempting to unite the twelve tribes from the diaspora by manufacturing the death of Jesus, the Romans united Rome with the death of Christians as a scapegoat. This tactic is as old as time and used today with 9/11, the Corona/Covid virus, and many more.

It's a sneaky way of getting rid of their enemies because they are too weak and cowardly to do it themselves. Now the Christians are persecuted and are killed in various ways of entertainment, all the while the Jews act as if they had nothing to do with it. They believe it's deflection of karmic consequence, but are too ignorant to understand it's building for the real punishment.

These Dark Occultists know the existence and operations of Natural Law and know they can't outright break it and get away without consequence. They believe they themselves are not actually performing the action, they commanded their mind-controlled slaves and cult members (order followers) to manipulate the Romans through fear and ignorance. They believe it shields them from the brunt of the karmic force, while the order followers inherit the "lion's share," as Aesop would describe, or the largest portion of the karmic debt. True Christians follow the conscience, and follow not orders.

The Christians are forced under ground.

Constantine's mother is how Christianity became the main religion in Rome; it only took 300 years. The Christian God helped him win battles. First they were tolerated, then it ruled even to a detrimental degree. Christianity spread through imperial households even though it was considered a slave religion.

Constantine replaced Roman Pagan ideology with Christian ideas. The 25th of December is the popular feast of the "birthday of the invincible Sun" which was the winter solstice now Christmas. Easter is an Anglo-Saxon name for the Passover which is the celebration of the escape from Egypt by the Jews, for Christians it meant the redemption of sin. Jesus dies and is resurrected on the Jewish passover, where Jews sacrifice lambs. The birthday of the Christian Church is on Shabuoth.

In 325 CE, Constantine would begin to convert Palestine into a Christian Holy Land. Constantine dies on the road to Damascus and the Pentecost making him the 13th Apostle.

The youngest of the family is Islam. Muhammad isn't the founder, he's the reciter of the dissertation from Gabriel. The word "Islam" means submission to God.

The Quran is one hundred and fourteen chapters, or more accurately Sûras, descending in length of revelations in the form of poetry given to Muhammad during his prophetic life. The Quran arrives from nowhere and instantaneously. Muhammad performs not miracles, he gives the Quran or Dissertation.

Muhammad was a marginal member of society, born around 570 AD and the revelations came around 610-632 upon his death. He was said to be an orphan raised by his uncle and is a part of the Banu Hashim tribe. He marries Kadijah and has many daughters. His wife was a capitalist in Mecca having a stake in trade and Muhammad was known as being a fair tradesman.

On a retreat, Muhammad is visited by Gabriel and is instructed to recite. Afraid, Muhammad says what shall I recite? Gabriel says again, recite. Muhammad says he has nothing to recite. Recite in the name of the Compassionate and Merciful, says Gabriel, and the Dissertation spills out.

Muhammad was overcome with fear for years, later he recites in public verbatim. The message was monotheism, the one God.

What came from Muhammads' mouth had nothing to do with him, but God only. Muhammad isn't a redeemer of humans, doesn't provide miracles, but is a vessel providing the word.

Some revelations were given in Mecca, the last in Medina. The main message from the Quran is, "There is no God, but the God." It presents a moral consequence of not paying attention to the word and an accounting of God in the last days. The good go to the Garden of Eden and the bad to Gehenna (a garbage dump outside of Jerusalem where there's a never ending garbage fire, believed to be the gateway to the underworld or Hell).

The revelations in Mecca were to convert the pagan religions (worshiping idols at the Kaaba) to the one God. The Medina revelation were for people already Muslim, and was instructional. Muhammad is barely mentioned in the Quran and is only the vessel from which the dissertation derives.

Yahweh and Allah are the same, Christians and Jews worship the same God as the Muslims, Allah means "God the Father." Muslims are Jews, and Jews are Christians.

The Bedouin nomadic tribe would go to Mecca and trade and worship for a few weeks to pagan shrines. The Quraysh was the main inhabitors of Mecca and profited from the gathering to the pagan shrines. Religion is the business of Mecca.

The Quraysh hear the Dissertation and care not. The Quraysh ask if Muhammad's God is so great, why doesn't he get us out of Mecca? There was very little water in Mecca and surrounding mountains.

But the word spreads fast, now everyone is afraid monotheism will destroy the economy of Mecca—pagan religion. Muhammad says the idols have to go and opposition arises. Muhammad's wife and uncle die and now only has his daughters and followers.

It's obvious that Muhammad will be assassinated and at the last minute pilgrims from Medina arrive and persuade him to arbitrate for them in the oasis of Medina, that has an underground water source and date palms for agriculture industry.

It didn't take much persuasion, so Muhammad and followers snuck to Medina and escaped Mecca. They believed this holy man could settle their civil war. The start of the Muslim era begins on this migration in 622 CE.

There were two dominating tribes inhabiting Medina, both tribes consisted of Jews, Arabs, and Muslims. Jews once controlled the oasis, but are now patrons to the agriculture industry.

Space available and the means to support the newcomers had to be discovered. Muhammad was provided for, the followers lived with other Muslim families and it wasn't a happy arrangement.

They developed the Medina Accords, agreeing Muhammad was the ultimate leader and all disputes would be handled by Muhammad. The Medina Accords made Muhammad the all and everything from tax collector, prophet, president, judge, leader of the armies on the raids, performed disciplinary actions for crimes against Islam, politics and religion are as one. All would live in the community—Pagan, Jew, Muslim. Muhammad preaches he's the last of the prophets.

The Jews in Medina accept not Muhammad as a prophet but are fine with the political arrangement. This was the main reason the Medina Accord had future trouble.

There was a tone from all the religions that the Israelites are good and the Jews or Judeans are bad. In Medina, the Quran becomes darker towards the Jews.

Islam changes, now Muslims pray facing Mecca, or more precisely the Kaaba, instead of Jerusalem. Muslims want to separate themselves from Jews and

Jews pray towards Jerusalem. Instead of fasting during Yom Kippur, Muslims now fast during Ramadan.

The remedy from civil war is to attack caravans traveling close to Badr Wells; having a common enemy is an easy way to redirect humans. Muhammad and followers highjack the caravans and become rich men.

Violence is introduced into Islam's equation. Muhammad permits violence against persecution from the Jews or the Quraysh. War between Mecca and Medina was inevitable. This was the first time two cities fought, before, it was tribe against tribe.

As time passes religion and politics become more of one and the same. The Jews and pagans in Medina become more unsettled and denied Muhammad as a prophet. So the first tribe of Jews or pagans are expelled, then another tribe is enslaved, later another tribe is executed on the spot.

Jews from outside Medina could trade without pressure from the political arrangement of the Medinanese. The Jews in Medina were suspected of treason of the Medina Accords and rightfully so.

Muhammad sends out raiding parties all over western Arabia. Overcoming all tribes around with terms of taxes to the new Muslim sovereignty. If a tribe is pagan, they will become Muslim (one who submits to God) or die. If they are Christians or Jew they will pay taxes and worship as they choose, they are already monotheist.

Remember, Muhammad is preaching monotheism and had a Covenant with Christians and Jews from then to today, it's religious law. Muhammad sent out groups to destroy pagan idols and convert pagans by force.

Muslims are required to pray five times a day, Friday noon prayer must be done in community. The first mosque is in Muhammad's house and the religion of Abraham is revived. Muslims are not required to pay taxes but are required to tithe a percentage to the Muslim community.

Muslim and Jewish women have the same restrictions of modesty and public circulation, but Muslim women have a pass from religious requirements. Muslim women can observe religious requirement but are not encouraged and have to be accompanied by men.

Islam was a political and religious enterprise at the same time. The idea of the separation of church and state is foreign to Muslims. Iran is an Islamic state with church and state simultaneously, religion directs government policy which causes humans to remain unfree to even learn lessons from God because man is their new God.

Muslims ask why not become Jewish or Christian? Muhammad says the Jewish religion starts with Moses, the Christian religion starts with Jesus, being Muslim is the original form of religion. The true Abrahamic religion and Abraham is the first Muslim or submitter to God. Muhammad circled behind Jews and Christians to take control of monotheism. It seemed Islam was going to take over as the main religion, but Christianity was too powerful.

Muhammad dies and everyone is surprised. They knew he wasn't God, but are still in disbelief. Muhammad made no provisions for his successor so the Muslims (the majority being Sunni) chose Abu Bakr to be the executive head of the community, not a prophet. There was a disagreement because people thought the appointment should have gone to Ali (Muhammad's cousin and son-in-law) and Ali's male descendants. The Imam (Muslim Pope) guided politically and was infallible.

The Shi'ites didn't like the idea of an Imam and understood that the Imam went into hiding and wouldn't return until the End Time. The Imam rules through a surrogate like Ayatollah Khomeini, the Father of the Islamic Republic of Iran.

Very little is known about Muhammad and the Quran mentions not of him. Muslim tradition thought it remembered, but their recollection is from the eighth century (750 CE) Baghdad. At this point it's of legend, Western Arabia is a land without a past.

Christians, Jews, and Muslims all think they have the word of God in written form. Jews have the Torah and believe not the New Testament is the word of God nor the Quran. The Jews believe not that the truth was revealed to the Christians or Muslims. Jews disregard Jesus or Muhammad as a prophet or Messiah. Jews believe the Bible must be understood literally, the Pharisees persist.

In the Torah, Exodus and Deuteronomy tells them to keep a record of the word on your arm or forehead, maybe instead of forehead it means before your eyes. The Jews take this literally and have little black prayer boxes on their arm and forehead called a Tfillin. Reminds me of the mark of man or the beast which is man.

Today the rabbinic definition of Jew is to be born of a Jewish mother and the right to return to Israel is the same. They are still Jewish even if they never follow the teachings of Judaism. You must state you want to be a Christian or Muslim and you can affirm an infant's belief into Islam or Christianity.

All three religions are reports of their prophet's sayings, changed, and sculpted society with gossip. Sharia Law is description, instruction, and gossip or consensus of society. All three take consensus or opinion of "experts" or lawyers discussing what is what with others, this determines behavior. Islam is a community based on consensus.

Doctrines develop in two ways—consensus or competent authority. Jews and Sunni Muslims are not only very much alike, products of consensus with a haze around the view. The similarities of Christians and Shi'ite Muslims are possession of authority—one with a bishop of Rome with a claimed prerogative for himself and parallels itself in the Imams who can pronounce dogma infallibly or through a later surrogate.

Christians and Muslims developed a doctrine of —the only reason for war is the defense of faith. The Christians have Missionaries, the Muslims have military forces. Islamic lawyers are reluctant to issue "decrees" or "enactments," while the Jews are in love with mandating religious and social conduct.

The Pharisees determined holiness of the Jews, and say to be a Jew is to be holy. Being birthed from a Judaean mother will not determine your Jewness. Most of the Pharisee's laws were about abstinence from foods and unholy people like the Gentile or nonobservant Jews. The Pharisees tried to program holiness with ritual purity and the Christians became more Gentilized, repression breeds excessiveness.

In the second century CE, we started to see spiritual succession from churches claiming to be founded by Apostles. At this point in this read you should already know linear succession is against freedom and the Lord. The absolute teaching authority of the bishops and council started to constitute living traditions.

Christians believe the Bible should be read figuratively, and is the word of God that would not exist without the existence of the Jews and the Torah. Jews celebrate the Passover according to the lunar calendar, Christians by the solar one. All off springs of the Jewish religion want to separate themselves from the Jews, which were reform Jews. Christians could have annihilated the Jews many times over but did not because they are committed to the Bible or Torah.

Christians couldn't demonstrate their validity without the Jewish Torah but started to by calling it the Old Testament. The New Testament now supersedes the Old Testament because it is now the charter of uplifting redemption. Christians believe the Bible is a document foreshadowing Jesus' life and allegorize the ritual and behaviors of the Old Testament. Baptism is central to membership to the Christian community.

It's indicated in the Gospels that Jesus offered a private and more detailed explanation of his message to his inner circle of apostles. These traditions and instructions were regarded as Apostolic Traditions. If these traditions are anything like the Talmud it will diminish the word. Depending on this unknown tradition, we would know the tampering of the word did occur.

Separate reform movements by the Christians and Jews occurred in an attempt to separate tradition from scripture. Protestant groups rejected the ideas that had no warrant to the scripture. Jews rejected the Mishna and Talmud as the works of man, not God. The words are evil men justifying their actions.

Christians say Muslims and the Quran is not the word of God. Muslims say the Torah and the New Testament is the word of God and the Quran is the third.

Jesus taught on his own authority and fulfilled the Biblical prophecy of Isaiah. Christianity could not exist without the Jewish Bible. Christians have the

Bible but are not determined by its guideline, the life comes from the New Testament. Christians call both parts (the Torah and the New Testament) the Bible and adhere to the Ten Commandments but not the dietary laws or anything else.

The Quran was an oral revelation from Muhammad's mouth, then the development of a current standard version was written 22 years after the death of Muhammad (650 CE) and the Quran itself admits it is ambiguous in places but must be taken literally. The Quran says itself some verses are abrogated or canceled and some are not, but define not which. This leads Muslims to believe the latter version from Medina is the new verses that must be followed.

For instance, the beginning Sûras from Mecca say there is no coercion in religion, then later in the Medina Sûras says find the unbelievers and destroy them—meaning Pagans.

Muslims recognize the authenticity of the Torah and the Gospels, but believe the Quran supersedes both. Muslims believe both the Torah and the New Testament have been tampered with by the Jews and Christians.

The Bible and New Testament were written around 40 years after Jesus' death, then re-written 300 years after Jesus' death. Constantine would canonize the new versions in 325 CE with the Council of Nicea. After 380 CE, Rome would conglomerate politics and religion with creeds and then dogma, this led to the Imperial Roman Catholic Church. Rome itself knew how to highjack the movement like the Christians and Muslims did, it too would do anything to survive. Christianity is far more institutionalized than Islam and Judaism.

Rome cares not of ideas or God, it cares about power. Religion together with government was the start of Inquisitions or investigations of heresy. Inquisition is from Roman Law to prosecute heresy. This empowered the magistrate to investigate and prosecute alleged offenders.

Muslims held the Holy Land and it seemed wrong to Christians. After all, this is where Jesus died and rose—among other sites. The Church convinces the monarchy to liberate the land from the Muslims, the Church couldn't declare war (the Lion's Share of karmic debt). We call it the Crusades, which were a failure. There were many Crusades, none of which were successful.

The Muslims drove the Christians into the sea, and there were no Christians in the "Holy Land" until Napoleon led his troops into Egypt in 1798.

Abraham, Noah, and Moses are central to all three books, Jesus is central to one (New Testament), Muhammad is central to one (Quran). The story of the Covenant is in two books—the Torah or Old Testament and the Quran.

All versions would be recorded and accounted for, but only a few were considered the word of God, the rest were considered the works of man. All books were eventually made into a standard version, all books say some of it is allegorical some of it is literal.

All are committed to God's transcendence, all believe true believers will see God and remain or be removed from God's presence. All agree on a Final Judgment in the Valley of Kedron (between Jerusalem and Mount Olivet) where many were buried from all three religions. The dead will rise and all will be judged. All believe everyone in the world should be their one true faith, all grew up in each other's shadow.

Revelation is God reaching down to his creation. Heeding God is to become a part of God. The interpretation of such has become a problem through lawyers, study, translation, and consensus. Those who claim to be more in contact with God claim authority, Jesus humbly diminished that bullshit.

Today one of these religions has deeply imbedded itself with a super power and is wielding indirectly its power. This religion is the "Israelites" directing the ignorant United States of America. The drivers of America have not the intellect to understand where the root source of their ideology derives, nor the understanding of the authority that directs them through ignorance and greed.

America is a Christian society by force. In the beginning of our great country, it forced everyone to be or become Christian. Yet it adheres to a Jewish authority because they are too ignorant to understand that Jesus separated us from the Jewish religion. They still believe that Jewish authority is closer to God than they, that they are a more pure religion because they're the older brother. All while claiming to be followers of Jesus.

God promised not the Israelites Jerusalem. That's a new idea from Jacob's descendants—the sons of a liar and thief, those who dishonors thy mother and father. The book of Kings isn't in the first five books of Moses, known as the Torah.

The Bible defines not the Promised Land precisely or consistently. The Israelites took what they could, until they were taken. The divinely-sanctioned boundaries in Palestine is a 20th century idea, not to mention the Fourth Geneva Convention Article 49 : Settlements are illegal and it was signed by Israel in 1951.

It's self declared and an American induced "Jewish State." Without the sustenance from the American citizenry, it couldn't survive. In this reality things that can't survive on their own perish. Not even Moses could tell anyone where the promised land was and he was in direct contact with God.

The Jew's lineage comes from a stolen birthright, it's the second son that inherits with intervention from greed or jealousy. However, the Christians are the second son to the Jews in religion, but this time the Jews want to keep it the first son, pretty inconsistent.

After Noah, Abraham, Moses, and finally Jesus, it is obvious to God that these people can't be helped, which is why he abandoned them. God had to exterminate the first group, have conquered the second group, and even just send a message to the third group and they still can't get it.

God continues to scatter the Jews like when he exiled them into Babylon. Neither God nor society wants not the Jews to live a David and Solomon existence. It's evil exceeded and was truly a dark time for humanity. God allows not for the Twelve tribes to re-unite, ten of the twelve tribes disappear for a good reason.

This David/Solomon experiment failed. These meat vehicles we inhabit are weak, the flesh is weak. Weak flesh is why the Jews are scattered, being human is to have weak flesh. Living in this existence is for learning, and some are slow learners.

Jews are chosen to lived under others sovereignty. God has tested these waters many times and know it must be so.

When Rome fell, the soldiers from the outer regions were abandoned by Rome and their conquered subjects. It was very hard for them to survive without the help of an army. Some starved to death, others would steal and trick their way through life for survival.

These soldiers and their families became Gypsies who performed street acts and became thieves and swindlers for survival. No one wanted these ex-Roman soldiers living in their town because of the way Rome treated the world. So the ex-Roman soldiers became nomadic Gypsies, to roam is to be like a Roman nomad. If they were Arabic they would be known as Hebrew, to wander.

America now pays for their sovereignty because of the Holocaust and psychological training of the American citizenry. Consensus is the only power of Israel.

The Babylonian Talmud became, for the Jews, the authoritative understanding of God's revealed Law. Many Jews understood this is from the hands of evil men justifying their actions.

What an evil thing to do to humans like genocide them in the Holocaust or any genocide for that matter. But do you really understand the Holocaust? Or are white people just in love with World War II because of movies?

It was the Gentile Jews from the Diaspora that Hitler was cleansing from the pure race, ordered by the Zionist Jews who believe themselves to be a more pure substance. It was also test subjects to see how much cyanide gas a human could survive—in the name of science. The 1924 Transfer Agreement was a pact by the Third Reich and Zionists to move Zionists Jews and their assets to Palestine before the genocide.

It was Zionists Jews that initiated the Holocaust, NAZI is an acronym for National Socialist (NA) Zionist (ZI). Together they were cleansing Europe of the Gentile Jews from the Diaspora. Hitler was a tool of these Zionists, and the Zionists were never considered to be the culprit. After all, they are Jews and Jews are being killed.

There was a German prison camp on highway 62, in Orange, Texas. It reside across the street from Godwin road. This "prison camp" housed German solider supposedly captured during World War II. However, these NAZI soldiers were free to roam as they choose. They could go into town, the movies, the grocery store. This sound like a prison camp to you? Or was this a way to infiltrate society with NAZI ideology?

In 1924, Count Richard von Coudenhove-Kalergi was the founder and president for forty nine years of the Pan-European Union, the forerunner of the EU. In his 1924 book, "Praktischer Idealismus" he says : "The man of the future will be mixed - raced... The European-Negroid race of the Future... will replace the diversity of peoples... Russian (Jewish) Bolshevism constitutes a decisive step towards this purpose, where a small group of Communist spiritual aristocrats govern the country... The General Staff of both... are recruited from Europe's spiritual leader race, the Jews."

This doesn't sit well with people who believe they are Jacob's descendants. Claimed Jew or not, they will not be the inheritors of the Covenant, diversity isn't on the menu.

The Jews from the Diaspora, who were Hellenized or mixed with other races in eastern Europe, were true victims of genocide from Jews who thought of themselves as the bloodline of Abraham or, more precisely, Jacob.

In 1950, Coudenhove-Kalergi was awarded the very first Charlemagne prize. Of mixed race (Asian and European) himself, two of his three wives were Jews and their lineage is from the Diaspora.

Zionists are master manipulators of the mind and masters of propaganda, one of the first is the book of Maccabees. Zionists view these actions as the ultimate fight to the end times they are so in love with.

Inducement of end times by the Tribes of Israel unite not, the only way you'll know the end has happened is after the fact. Continuing to feebly attempt inducement will prolong what they desire. What we resist, persist. The Zionist's controlled the Third Reich to receive the Lion's share of the Karmic debt and all of the heat from the world, all of the attention is on the NAZIs.

Zionists desire the City of David (Mount Zion) and people from Jacobs lineage only, all other humans are Goyim and must be killed. Yet they can't survive on their own and are constantly being conquered, it takes $10 million US dollars a day to keep these people afloat. Zionism is the anti-Semitic separation of Davidic Jews from Gentiles, Jews, and Arabs.

Being Jew has nothing to do with being a descendant of the House of Jacob (Israel) or the House of David. Being Jew is destinational to Judaea, Jesus was from Galilee.

Zionists are drunk with ignorance. It has over-powered them—now they are vomiting it up. Empty yourself of darkness- you will be filled with light.

The world is waking up to the understanding that Israel is an apartheid state. Israel is genociding the Palestinians with actions like befouling the Palestinian's water supply, spraying their homes with raw sewage and bio-hazardous waste, and moving them away from the coast.

Now the Palestinians live in an open air prison (reservation)—unable to leave by land, air, or sea. No food or clean water—medical supplies air dropped by sympathetic countries, and they are slowly being squeezed to extinction. Israel is breaking many international laws like Fourth Geneva Convention Article 49.

"It would be my greatest sadness to see Zionists do to Palestinian Arab much of what NAZIs did to Jews." - Albert Einstein

Benjamin Netanyahu says, "Israel isn't a state of all its citizens… Israel is the nation-state of the Jewish people—and it alone." That means every other human isn't as pure as they. This is how the Holocaust developed.

Even Kuwait's Parliament passed legislation on boycott of Israel, banning normalization of ties. Only 1.6% of Americans owned slaves, most of which were Zionists. Hexagram is the Seal of Solomon rebranded as Star of David.

This Locust trait must be weeded out and eradicated from our existence. They'll try to convince you they are pure blood and you are Goy. Zionists are just as less than as everyone else, even more less than. The National Socialist Zionists are the NAZIs.

These people who claim to be Jews, who are not, bear the mark of Cain. Just as Romulus killed Remus for Rome, Cain killed Able and Cain's descendant (Enoch) built a city. But where did they go?

In 740 A.D., land locked between the Black and Caspian Sea was a land known as Khazaria, which is now Georgia, and parts of Russia, Poland, Lithuania, Hungary, and Romania. It was said to be the birth place of the modern Jews from the Diaspora who are not Jews. The Khazars had Christians closing in on one side and Muslims on the other and feared attacks from both.

The Khazars practiced Ideology Pagans worship with, much like Solomon's Temple, such as human sacrifice, and were neither Christian nor Muslim. This was a time of conversion, you can be one or the other, and there is no middle ground.

The Khazars sacrificed horses to the sky god Tengri, the supreme deity. Sun amulets were widespread as cultic ornaments, a tree cult was also maintained, and they many times worshipped ancestors.

The Khazars believed death was a continuation on earth and crossed through the divide with weapons, food, horses, and servants or soldiers. A funeral of a Tudrun around 711 AD sacrificed 300 soldiers to bring with him to the underworld.

King Bulan understood to protect themselves they would have to convert to Christianity or Islam. The king also knew the Jew could handle both sides along with trade. So the Khazars adopted a fasade of Judaism by the greater part of the ruling class in about 740 AD, having converted to Rabbinic Judaism. The Khazars and the Greeks are where the European and Russian Jews derives, and is the connection to the Ashkenazi Jew with the Yiddish language.

Many like the Khazars would be taken over by religion while continuing past religious practices underground.

The Khazars seemingly converted into Islam as well. During the 10th century, a Muslim geographer named al-Istakhrī claimed the White Khazars were strikingly handsome with reddish hair, white complexion, and had blue eyes. Muslims dye their hair and beards red as well. They do this because it is said that Muhammad had red hair and beard. Was Muhammad a Khazar?

Muhammad came from nowhere and was adopted by his "Uncle." There is no clarification of his parents. Could it be that Muhammad was psychologically conditioned by the other 10 tribes of Jacob to regurgitate the Dissertation unknown to himself to regain control of the Abrahamic Covenant?

These people may very well be the other 10 tribes of Israel. Abraham's patriarchal family were

Pagan Idol worshippers and practiced human sacrifice, like when Abraham was to sacrifice Issac. There is always someone leveraging for control of the mind.

Byzantine emperors Justinian II and Constantine V both had Khazar wives. These Pagan wives influenced these emperors just as David's and Solomon's Pagan wives influenced them.

Pagan worship is how darkness entered into the church. The elite aristocratic clergy is how the Abrahamic, Christian, and Islamic religions became violent and dark forces pitted one against the other. The slow infiltration for control of the human mind is desired, and self eradication is even better.

These powers are greatly afraid, their fear of death causes them to harm others. We must free them from their pain.

People of Pharisaic existence are people who can only think with their left brain and have not the ability to think with their whole being. I believe it's because they are a faded copy of what a complete human should be. These Pharisees only have the ability to follow orders, thinking for themselves isn't even considered, they haven't the ability.

It is the Pharisees, Scribes, and Chief Priests who found fault in unwashed hands, it was against the laws and traditions of the elders. Pharisees love laws and know not why, they just love them. Hypocrites!

They would rather wash thy hands than love thy neighbor. Jesus made the blind see, but it happened to be the sabbath. The Pharisees threw rocks at him, but some murmured how can a man who does such a miracle be in sin? There became a divide between them because of an internal struggle between prescribed indoctrination and what their entire being tells them.

The Pharisees loved the praise of men more than the praise of God and have no cloak for their sins. They simply know not God.

Even the disciples of Jesus couldn't think for themselves, long indoctrination prohibited their ability. Jesus tells them to take heed and beware of the leaven of the Pharisees and of the Sadducees. They speak among themselves, "Don't eat their bread?" They haven't the ability to connect fully and think with their entire being. The disciple can only focus on the words spoken but didn't pick up what Jesus was dropping off.

Jesus finally says, "How is it that ye do not understand that I spake it not to you concerning bread...", they finally understood he was speaking of the doctrine of the Pharisees and of the Sadducees. What a painful existence for Jesus, to have to endure such ignorance from all angles and constantly clarify word meaning for humans that are too lazy to know in their being, I feel for Jesus. What a painful experience worse than death.

Why did the Jews plot to kill Jesus? Because he performed miracles that made men believe on him and the Romans would take the Jew's place and nation. The Jew's twist was Jesus should die for the nation and gather the children of God that are scattered. They were going to put Lazarus back to death so no one would believe on Jesus. What they didn't know was he indeed was going to die and destroy their plan along with it, and even today the Jews gnash their teeth.

Jesus said he is the porter and the door way. "I am the door: by me if any man enter in, he shall be saved, and shall go <u>in and out</u>, and find pasture." Death is not the end because you can go in and out of this existence. The sons of man have the power to lay our life down and take it again. The Good Shepherd doesn't see a wolf and flee, he lays down his life for them.

The source is all and we are I Am. These Babylonian gods and the Pagan gods of Abraham want to separate you from the source and destroy the earth by fire. There is nothing separate in this reality. We are not many, that's absurd. There is only the oneness that subsumes the parts. The Lord and the wind is our next of kin—that which nourishes us like mother's milk. If Jesus saved us from anything it was the failed and barbarous Jewish religion.

There is a limit of doing harm as you choose, large or small, and later asking for forgiveness. The disciples ask Jesus how many times must we forgive our neighbor? Jesus replied 70 x 7, this is also the forgiveness diet where you can write 70 times for seven days in a row, "I John Doe forgive you Jane Doe."

Without missing a day, by the end of the seven days, you and the person you forgive will change. If you complete this number you will find true forgiveness, if you find not forgiveness—do it again. Without forgiveness you indeed will live a hell on earth. We are all God having an experience with itself simultaneously. We are a piece of God like a rain drop of water in the ocean.

Conclusion

XXVII

The Kingdom of Saudi and their accomplices have exceeded many in cruelty, theft, and murder.

American oil men thought they would drain the Kingdom of Saudi of their resources and would become sheikhs and kings, but the joke is on them. You will die of respiratory infections before that ever happens.

The oil in Saudi is the same oil in America. You can deplete not a resource on the planet without destroying the entire planet, its called seepage. It would be like taking blood from your arm and believing you can steal it from you leg. The entire being will perish. The possessions of this life are too pleasant, and such pleasures grip the soul by the throat, holding it down to earth. Our possessions possess us.

When we break our oil habit—we can get rid of the Intercostal Canal and regain our wetlands to protect us from powerful storms. Y'all took away the wetlands—and now Mother Earth is making a new one. The swamp will now be where your house is, you better get your ass up on stilts cause the swamp is coming.

We need not the oil business, we need high paying jobs that truly benefit humans or free energy to help keep us fed, warm or cool, and the ability to travel with ease. We need not the oil business, we need ease in survival. We need not the oil business, we need to love thy neighbor.

The pipeline everyone lusted for or any new pipeline would benefit not anyone but corporate communist Canadian oil men, and the pipeline owner. Refining Canadian oil and sending it back doesn't mean you're exporting oil—it means you're the polluted white trash that lives with the consequences of refining it.

It would produce not more jobs. Our refineries run at 100% at all times, and are more efficient on the weekends when the engineers and asshole managers are away. There is oil all over the planet, even under the refinery. Most all refineries belong to the Kingdom of Saudi, and the Kingdom of Saudi, along with the United States government, decides who they will allow to become rich.

You can operate not a Hydrocracker without platinum. The United States government owns all the platinum used in refining, and leases it to the refineries. It then reclaims it, refurbishes it, and rents it out again. The United States government can make or break any refinery it chooses. You and I couldn't refine oil unless the United States government allows it to be so.

Just because Canada lacks freedom of speech and a great deal more freedoms advanced societies posses doesn't give America the right to pollute them. Yes, they have communist characteristics—Chinese people have the right to clean air and water as well.

Oil produces mobility and energy for our daily lives, we need a good stepping stone before we cut dry.

Separating hydrogen from water is too small of a step, and will also leaving one hydrogen molecule exhausting from the separation. This remaining hydrogen molecule will attach to hydrogen in our breathing air causing it to become O3 (dangerous Ozone). Possibly causing more harm than oil combustion. If you have to create an explosion to produce energy, than you are a fucking caveman. We need new inventions that are beneficial to all living beings.

We will need natural gas for a time, the February 2021 freeze made that very understandable. We can survive not with coal plants, most every human is at a threshold of pollution exposure and can receive not anymore pollution without experiencing severe illness or death from what some rebrand as Covid. They have made it impossible to labor outside without respiratory infection from pollution, mostly from Africa and Mexico.

Rolling blackouts in Texas is bullshit and third world. Austin, and Texas in general, has too many cows per acre and not enough electricity to cover it. For you ignorant city assholes, that means Texas is over populated. Server farms, WIFI, and many other products that lack beneficial qualities are draining our electricity, Earth's resources, and our very being. We have lived a long time without these products, we can make it much further without them.

This is why Los Angeles has a dangerous amount of carbon monoxide surface area that is killing everyone in the vicinity. Electricity comes from a generator of electricity—from oil and diesel.

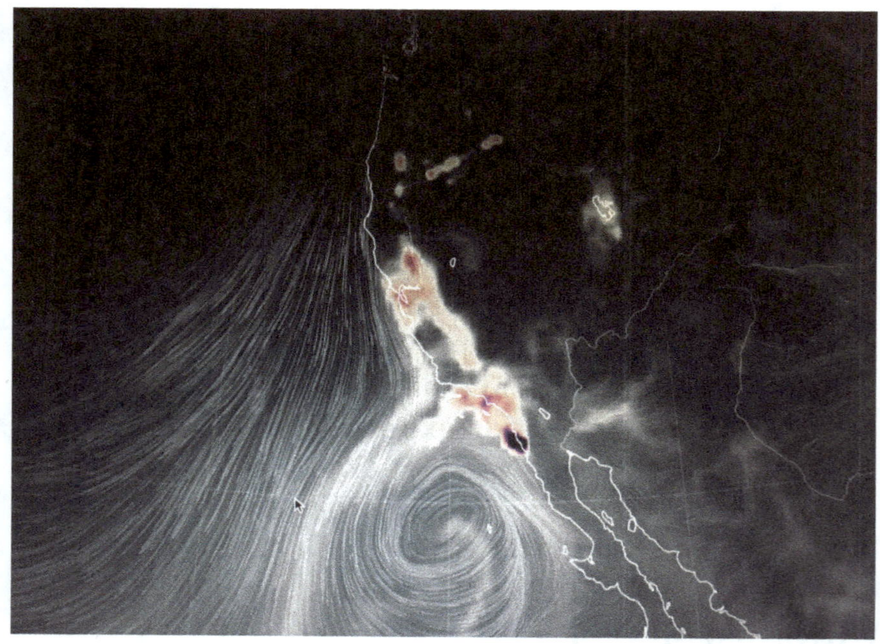

Believe not Elon Musk that civilization will crumble because of a dwindling population—the population of the Earth is ramped. How can you trust someone who doesn't like to fuck? Just because someone is rich and goggleyed doesn't make them all knowing—take your ass back to South Africa. Texas doesn't need a stronger energy source - it just needs to suit a population of what the land and water source can tolerate, it's called balance. Everyone else go somewhere else—Elon.

Our infrastructure is piss poor—dams, bridges, electrical systems. It is unacceptable to have rolling blackouts for any fuckin' reason.

The governor of Texas could have salted and sanded the roads to allow Texans to travel during the freeze, but didn't do so. Greg Abbott wanted you to freeze your ass off so you would beg for a dumpster fire to create energy. Greg Abbott is like an abusive husband who will harm you to get what he wants, which is most likely pay-offs from industries or the communist party. People died in the freeze, let's not forget that.

Do we need a Democrat in the Texas governors seat? Fuck no! They'll have Kowalski from Ren and Stimpy playing on the girls high school volleyball team.

Now we have an election coming—we can chose from an openly admitted communist (Robert Francis "Beto" O'Rourke) from El Paso, who in September of 2019 said on national television he is coming for our guns. O'Rourke can't take care of his own town, has several blond Russian speaking women at his dispose who apparently came to Austin in a "cell" or group with fifteen hundred others Russians as Obama dreamers—and the hidden communist (Greg Abbott) in the seat now. Strange times in Texas.

If we need to invade New York City and Los Angeles to rid them of communism so the people who over populate Texas can return home, then so be it. We will do what we need to restore balance. If you don't want to freeze your ass off, head down to Mexico and fight the cartel. Mexicans seem to love the American way, let's give it to them.

In no way should we worship or be loyal to an industry like oil or any company. The only industry or company that matters is the one that is truly beneficial to you and me. The citizenry in the companies are more important than industry, company, or refinery. When the industry, company, or refinery lack beneficial qualities, we must beat it with a shovel and bury it behind the barn. We can leave an industry and pick up a new one with ease, they are a dime a dozen.

You must understand what team you're on and identify the enemy. Exxonmobil isn't for America or our employed citizens. It belongs to the Kingdom of Saudi and the Kingdom of Saudi does as it chooses with its belongings. We must take care of our citizens that work inside of Exxonmobil and strong arm the company itself.

Exxonmobil is now in a lockout of it's employees and trying to recruit unsuspecting collage students a Lamar University—that Exxonmobil owns. Young people of Southeast Texas, hear me for those who can hear, Exxonmobil will brainwash you and cause you great mental anguish while causing you to commit crimes against the World and Universe. I know you're tired of double shifts and being a full time student—there is a reason why we call people who cross a picket line a Scab.

They can leave not and take the plant with them. They can try to remove their program and pay, we don't need it. We'll just take their tangible assets, do it ourselves, and laugh our asses off. With any luck they will retaliate with violence, giving us universal authority to do as we choose in protection of our person and property.

We have the ability to replace the substance referred to as fossil fuel (the coolant system and blood source of Mother Earth), the question is do we have the courage to replace so called fossil fuel? Or would we rather destroy the earth by fire?

Capitalism has become a predatory stance. Capitalism looks to be an agent that seeks creativity and energy—then dark forces steal it and parasitically take its place. Creativity must be guarded from greed.

The Federal Reserve, which is neither federal or a reserve, along with the United States government can build or break the economy as they choose. It is only the appearance of Capitalism, true Capitalism flourishes not under a banking cartel.

Humans are continually conditioned to trade promissory notes for other' promissory notes. Now we trade paper notes (dollars) for digital notes (Bitcoin). The digital note is more fiat than the paper promissory notes, we can actually touch the paper promissory note, buy groceries, physically hand it to one another, and it's in your back pocket.

You need not electricity to receive it, there are too many obstacles and middlemen between you and digital notes. Not to mention bitcoins are paid to infiltrate the human mind, turn wants into needs, to know and sculpt your desires.

Worry not, humans are different people every seven years—even cellularly. Once we throw down this make believe bullshit called the cyber world, the controlling forces will lose their grip as a short time passes. Your wants and desires will change and the leverage used against you will dissolve.

The hackers that hacked the Colonial Pipeline extorted or stole millions in bitcoins from Colonial; later the FBI seized those bitcoins from the hackers. If you believe that any currency is cryptic then you have your head too far up your ass, so far that you can't hear or see, and when you speak only shit comes out —you have shit for brains.

We are designed not to worship fiat money or trade our souls for paper or light numbers in a make-believe world behind a screen, we are here to progress with the unmoved mover that moves in all that moves. We are the source having an experience with itself simultaneously—I Am you, and you are I Am. Capitalism is a great way to train people to become psychopaths.

Psychopaths care not of others' health or well-being, they just want to win—money, power, food, shelter, sex, control, ease.

We make laws and beautiful buildings to give the appearance of civilization. Civilization is a fasade, and every human inhabits an animal. Most of the time the animal we inhabit (your meat vehicle or conscious corpse) directs the soul which is disastrous. Your car doesn't tell you where to go, you direct the car, and so it should be with your flesh. Your soul is the drive, you tell your flesh where to go. The line that separates civilization and the savage animal we inhabit is thin. I dare to say animal, animals have not wars.

And what is it that separates us from these beings that speak not english? Is it our ability to trash and leave the planet behind? Why do all planets in our solar system have condensed brine? Most likely from the draining of their oceans. It was removed planetarily or evaporated from the heat of its core. The planets lost their coolant system and over heated.

Now we have billionaires building ships that leave our atmosphere and gravity for a short while. So far Virgin Galactic Flights bested everyone and we received a clear view of the flight, except for all the other daredevils that shot themselves into space before these billionaires and one free fell back to earth. Space X has been blowing up rockets in South Texas, destroying the water supply and everyone cheers them on.

If Space X is picked to build a ship set for the moon, then what the fuck do we need NASA for? You brain dead motherfuckers think NASA is the Government, they ain't. At this point all space exploration is circus and bread to keep you from looking at your troubles on earth. I don't care how many times you saw Star Trek, the evolution of humans isn't somewhere else—it's here.

These assholes want to trash the earth and leave the destruction to the minorities. You are Locusts—your train of though needs to be trashed and abandoned. Anyone who redirects our attention from our planet is the enemy of the planet and all living beings. There will be no white flight away from Earth—there's no where to go.

Just ask Gil Scott-Heron, whose sister Nell got bit by a rat, but Whitey's is on the moon. We must stop all travel into space until you have mowed the lawn, take the trash out, clean your room, feed your pet, do your homework. Then we will discuss whether or not you can play spaceman. We can still observe space from Earth, but must redirect our priorities. We don't even know what's in the oceans, and the Chinese are dredging and destroying it at a ramped pace.

If these scientists are so smart then why do people still hunger, and freeze or over heat? Why do children in Texas die of starvation? Why are humans living under overpasses? Why must we slave for products that satisfy not and destroy the Earth? I'm not asking you for an explanation, I'm demanding solutions as a priority.

The government subsidized Space X, along with Google, Amazon, and other industries that provide not any real benefit, are overpopulating Texas and bringing an element of human Texans despise, which can be identified as people who can think not for themselves, have more money than sense, and give up their freedom for safety.

We have nuts and fruits of our own, we need not a new variety. If their way of life was so good, why did they leave it? We have our own way of life that isn't perfect by any means. But it hasn't degraded to the extreme that these fleeers are fleeing. Continuing with this lifestyle is detrimental, and anyone who continues this detrimental way of life is insane.

In appearance the said breed of humans have occupied Austin and Central Texas. If they are smart they'll know Austin is still Texas, and Texans get wild from time to time. Texans are asleep at the moment, but the sun rise of truth is coming over the horizon. But keep coming and bring your tangible assets. You may find the gate you passed through closes behind you like a hog trap. Did you see a trail of corn leading up to the gate? Pay no attention to it, just keep following the corn.

Texans are realizing that we have become the Mexico of California and New York, with cheap taxes, land, and labor.

On November 17, 2021, the Austin American-Statesman (which is false advertisement to call themselves Statesmen) reported, "Cheaper housing and room to grow…" The infiltration and narrative of the Austin American-Statesman is a lie. I know Texas corporate lawyers who are out bidded on homes and property by Californians everyday.

The tech industry and their connection to the Communist party has allowed them to become very rich through state sponsored murder and slavery. The tech economy is designed to disrupt Capitalism and the American economy. They have overpriced homes all over the world—even in Newfoundland—a very small fishing island. They want to sell out, get rich, and buy a farm. Just buy a fucking farm before selling your soul.

It is very likely Texas will become a country again to salvage the ability to save room and resources for our Texan citizens. There is hardly a breed of American more blood thirsty than Texans, and we don't give a fuck about political correctness. There is a correction coming and there are ways you can alleviate your pain. If you're a refugee from a communist state in the United States and must move from your home for survival, then consider these ideas.

When in Rome, do as the Romans do. It's best to keep a low profile. Tearing down homes and replacing them with piles of condos is a no-no. Building modern monstrosities is a mistake, repugnant, and identifiable. Running for office to make it like where you came from is ignorant and obvious that your here for political ideology. Infiltrating our academic sector like collages and school boards is immediately dangerous to your health and well being.

Those ideas you want to implement are what caused you to degrade your former residents. We like wide open spaces, houses on top of houses anger Texans as bad as high rises. Remember, refugees refuge

and return home with help. Thank you for being polite, it carries a lot of weight, we simply haven't the
resources for you or your ignorance. We don't want a
bunch of fucking cowards in Texas, bring your ass
home and fight.

    Judges in San Diego struck down vaccine man-
dates because their inconsistent to State law. Counties
in New York will participate not in any mandate.
Everyone of authority will run for the hills. Soon a
mandate will return to mean two men went on a date.

There are tall buildings looming over the state capital, one of them is The Frost Bank another is the W Austin, and more. These buildings along with others appear to dominate and control the state capitol building. The view is telling everyone that money and power control the state capital, maybe it does.

A 1931 ordinance of a building's maximum height of 200 feet preserved the visual preeminence of the 303 foot Texas State Capitol. In the 1960s the Westgate Tower was built with 26 stories and angered Texans. In 1983, Lloyd Doggett wrote a bill protecting the view of the capital.

This has nothing to do with the view in Austin, this is about respect for Texas and over populating of a small water supply.

Growing up, I was told that no buildings would ever be taller than the Texas State Capitol. Now we have the ugliness of Houston and New York City. If there isn't a law preventing this mockery of the State of Texas, then there will be new laws implementing the 200' rule. We will have them tear down as many levels from their buildings as necessary at their cost. We will remove the operating license of many corporations Texas wide. Any building owned by the Communist Party/affiliates or the Kingdom of Saudi will be seized and demolished completely.

I must say that the people from California and New York are Americans. I know there are many good hardworking people from these states that must refuge to salvage what income or retirement they have because of over taxation without representation, cost of living, and communism.

They are also refugees from a massive amount of pollution from industry and wildfires. Californians and New Yorkers are our brothers and sisters, we must help them return home. You must slay your dragon and face your fears with ferocity and faith. We can help you remove communist authority with force if necessary. Refuging means you are running to receive help, we will help. You must get tougher than you were when you ran. You must have on your mind, "I may have to kill these psychopaths that destroy our freedoms and liberties."

We must refrain from making Austin weird again, we must make Austin Texas again. If that doesn't work Texans, we must inhabit their newly burned land and make it our own. Texans, remember what Doug Sahm and Waylon Jennings said, "That you just can't live in Texas, unless you got a lot of soul."

We are cutting it too close to the bone, a lot of people have been driven back to caravans (travel trailers and tiny houses). We are trying to make lemonade out of lemons.

Our land simply will survive not with the amount of people living on it, we will survive not with the concentration of consumption and waste emitted from these humans. We must live within the real limitations the Earth provides. We must find mutual salvation and remedy the root cause of the distress these humans suffer.

This means we will need to redirect where humans develop habitable structures for living and commerce. The land in Texas, especially central Texas, is already overpopulated and measures must be taken.

An easy way to weed out corporate psychopaths and find companies that truly benefit Texas is Covid and Communism. We will evict any company that understand not freedom of Texans.

We must take away the operating license of all companies that have ever require their employees to show proof of vaccination. If any company already committed tyranny towards any human in the matter of vaccination, employment, or entrance in to a public accommodation, they must be punished with meaningful, steep penalties and have their operating license removed and more.

Any company with communist characteristics must have their operating license removed. Any company or government agency partnering with communists, who enslave, murder, and control human consciousness must have their operating licenses removed or removed from office with fines and jail time immediately. Any company that censors content must have their operating license removed. We must seize all their assets for crimes against humanity and freedom.

The large companies that make billions must pay for the privilege of doing business in Texas. Giving billion dollar companies exemption from taxes for an extended period is bullshit. They have it and we must take it. New refineries and chemical plants have an obscene amount of money and must pay for the privilege of doing business in Texas. Creating jobs means nothing if our quality of life diminishes, and most every Texan can agree that our way of life is diminished.

We are living a reality were you are free but some are more free. To charge someone with a "hate crime" is saying some lives are more valuable than others.

Gays went from equal rights to supremacy. America celebrates black lives and queer lives—not the Natives (which include people who are perceived as Mexican). They celebrate people who don't know if they're a man, woman, or a toaster oven. Watch the episode of Mr. Rodgers where he explains that boy don't turn into girls, and the same for girls.

Ladies, how do you like these buck kicking your girlie ass in every sport—become a bricklayer, that'll show them. Caitlyn Jenner was a magazine woman of the year—first year to have tits and she beat every woman on the planet.

Black people are free—not the First Nations. They are invisible to you—a thing you learn from a history book or watch being shot as the evil savage in the movies. A thing to learn about under florescent lighting and forget.

The celebration of the mentally ill is the perpetuation of the illness. People who purchase sex changes never find their happiness and are very prone to suicide. They have a right to do as they choose with their life, they have not the right to force the issue on anyone else, friend or foe. I worry and have compassion for these people who have been manipulated to mutilate themselves and never find satisfaction.

Critical race theory is conflict theory, always the oppressed and oppressors. Being at a riot means not everyone at the scene has committed a crime. Individuals in mobs commit crimes by aggressing others' person and property and the whole mass of people are innocent until proven guilty. This applies to bigotry as well.

Being white doesn't make you prejudice, even if you were taught it at a young age or say offensive words. You have the right to think what you want, just as gays, Asians, and black people do as well. You have not the right to aggress on their person, property, or peace of mind. Actively sabotaging humans and premeditative action designed to aggress on others' person and property is against Universal Law with a steep penalty.

This and many other tactics are meant to divide us as not only Americans, but humans. Once we come together the true aggressors and government oppressors are finished.

Civilians have the right to be as fucked up as they choose, unless they aggressed and forced communism or anything involuntary. Thoughts are legal and they can think as they choose.

The January 6, 2021 "Insurrection" was a way to weed out the most willing to fight for freedom. Former President Trump helped them, he's a Democrat and was a pusher of the vaccination and the dangerous 5G telecommunications weapon. You Trump lovers have cognitive dissonance and are idol worshippers. You Trump lovers will defend him when I tell you he wanted everyone to be vaccinated, your reply is, "it was an emergency."

21 US Code, 360bbb-3 <u>No EMERGENCY AUTHORIZED MEDICAL PRODUCT CAN BE MANDATED</u>. e 1 A ii III "the option to accept or refuse administration of the product." Masks, The PCR Test, The Covid Vaccine are all Emergency Authorized.

Former President Trump is in violation of 21 US Code, 360bbb-3 and must be punished as an example to make very clear the rights of free humans given to us by a higher power to all humans at birth.

In a statement on Friday, April 2, 2021 by the Principal Deputy Assistant Attorney General for Civil Rights Leading a Coordinated Civil Rights Response to Coronavirus (COVID-19), Principal Deputy Assistant Attorney General for Civil Rights Pamela S. Karlan issued the following statement and attached resource guide to assist Federal agencies, state and local governments, and recipients of Federal financial assistance in addressing ongoing civil rights challenges related to the COVID-19 pandemic:

The COVID-19 pandemic has stressed our Nation's commitment to an open, equal, and inclusive society. We have seen hateful and xenophobic rhetoric and violence aimed at Asian American and Pacific Islander (AAPI) communities and businesses.

We have also seen Black, Indigenous, Latino, and Pacific Islander communities, as well as people with disabilities, suffer disproportionately high rates of death and greater risk of infection and hospitalization.

COVID-19 has magnified social, economic, and environmental inequalities that we cannot ignore. As a Nation, we cannot adequately respond to, and recover from, COVID-19 if we do not protect all of our neighbors. That requires us to pursue justice on behalf of those targeted because of their race, color, religion, national origin, sex (including sexual orientation and gender identity), disability, or citizenship.

The Department of Justice will vigorously enforce Federal civil rights as we continue the process of national reckoning, recovery, and healing. Civil rights protections and responsibilities still apply, even during emergencies. They cannot be waived.

Federal agencies, state and local governments, and recipients of Federal financial assistance are an integral part of our shared effort to uphold civil rights.

The vaccine mandate included not Postal workers, because their union said they'll shut down the mail—good for them. That's one sector with testicles.

21 US Code, 360bbb-3 <u>No **EMERGENCY AU-THORIZED MEDICAL PRODUCT CAN BE MAN-DATED**</u>. e 1 A ii III "the option to accept or refuse administration of the product." Masks, The PCR Test, The Covid Vaccine are all Emergency Authorized.

THE AMERICANS WITH DISABILITIES ACT. The ADA prohibits discrimination on the basis of disability. Title III: Public Accommodations "Public accommodations must comply with basic nondiscrimination requirements that prohibit exclusion, segregation, and unequal treatment."

The CDC states that a person who has trouble breathing, is unconscious, incapacitated, or otherwise unable to remove the face mask without assistance should not wear a face mask or cloth face covering.

Centers for Disease Control and Prevention (CDC). (2020, May 22). About Cloth Face Coverings. Retrieved June 2, 2020, from https://www.cdc.gov/ coronavirus/ 2019-ncov/ prevent-getting-sick/about-face-coverings.html

U.S. CIVIL RIGHTS PROTECTION

MY LEGAL RIGHT TO ENTER, SHOP AND BE SERVED AT THIS ESTABLISHMENT -- without covering my face or showing proof of vaccination -- IS PROTECTED BY STATE AND FEDERAL LAW

1. **This private business has a LEGAL CLASSIFI-CATION as a "public accommodation" according to title III Re 28 CFR 36.104. Your private business serves the public and therefore must abide by**

all state and federal laws. No business policy supersedes the law. No Governor's order, health order, emergency or pandemic supersedes constitutionally-protected rights. This business is open to the public, and I am the public. <u>Your denial of my service violates several federal laws.</u>

2. Federal law 28 CFR 36.202 prohibits "denial of participation" from this business establishment. 36.202(c) states that unless I have been individually assessed as a "direct threat" you may not exclude me from the SAME and EQUAL services as others.

3. Denying my service or requiring me to be served outside or be limited to home delivery is a VIOLATION of Title II,III, and VII of the U.S. Civil Rights Act of 1964.

4. Title III, Sections 36.202(a)(b)(c) and 36.203(a)(b)(c) states that I shall not be denied the same PARTICIPATION and EQUAL ACCESS as everyone else. The law prohibits you from serving me separately or differently.

5. As such, this business is PROHIBITED from unlawful discrimination by denying the entry of any member of the public who is not disturbing the peace. To do so is a crime of unlawful restraint and interfering with commerce and you will be held personally liable for this crime.

6. These premisses are open to the public and thus any charge of "trespass" is a false accusation as I an complying with all lawful conditions allowing

me to remain on these premises and be served by this business without discrimination. I do not need to disclose my condition to you.

7. Learn about your rights and how to defend them at www.TheHealthyAmerican.org

Direct Threat: Legal definition

There is no evidence that I am a direct threat to the health and safety of your business.

According to Title of the U.S. Civil Rights Act 36.208, "In determining whether an individual poses a direct threat to the health and safety of others, a public accommodation must make an individualized assessment, based on reasonable judgment that relies on current medical knowledge or the best available objective evidence."

Absent a court-order of quarantine or isolation, there is no medical evidence that I am a threat to the health and safety of your business. Innocent until proven guilty in the USA. Therefore, my rights to equal access to the goods, services, privileges and facility of this establishment is guaranteed by Titles II, III, and VI of the Civil Rights Act of 1964.

Arrest Warning: You are hereby notified that state and federal laws make it a crime to deny the Rights of an individual. <u>You can be arrested for this crime</u> and held personally liable for criminal and civil damages, including fines and jail time. That means you can be personally charged and arrested for this crime, regardless of what your manager, governor or health officer says. No law or store policy supersedes the Federal law.

You can learn more about your rights at www.TheHealthyAmerican.org.

This plan has been enacted many times before. The true handlers of the nefarious plan enacted over and over generationally waited for people who are fighters to become too old to do so. The new generation has been psychologically conditioned and chemically neutered to ensure success.

Trump ensured the willing fighters would get caught in a world of judicial shit to wear them down from the real fight, Trump is the False Prophet. This is why they let them into the capitol. I know people who were there and who work there, they let them into he capitol.

Now it is said the FBI has arrested many and are using cruel and unusual punishment on these "terrorists", and no one is permitted to speak with the detainees. As I said many times before, we have a right and <u>duty</u> to overthrow a government repugnant to the constitution.

The FBI has made itself an enemy of Americans. This has been happening sense the first time J. Edgar Hoover squeezed his fat ass into a pair of pantyhose as a cross dresser. The very founding of the FBI has been political, and they have killed many leaders foreign and domestic. The FBI's very existence is for political leverage.

We can allow not for the FBI to kidnap our citizens one by one, hold them in undisclosed areas or use cruel and unusual punishments. We must protect and save our brothers and sisters. George Bush and 9/11 ensured the removal of our freedoms and the acceptance of these devilish means, we must rectify these mistakes immediately.

The FBI has performed a raid on journalist homes that work at Project Veritas in search of a diary that belongs to president Joe Biden's daughter that was not printed and handed over to authorities because of the lack of authentication.

Now a federal judge orders a special master over the seizing of Veritas phones and cites "journalistic privilege." We must see the affidavit submitted to get the search warrant.

On November 14, 2021, Brian Hauss, senior staff attorney with the ACLU's Speech, Privacy, and Technology Project, released a comment on the FBI raid of Project Veritas stating: *"the precedent set in the case could have serious consequences for press freedom. Unless the government had good reason to believe that Project Veritas employees were directly involved in the criminal theft of the diary, it should not have subjected them to invasive searches and seizures. We urge the court to appoint a special master to ensure that law enforcement officers review only those materials that were lawfully seized and that are directly relevant to a legitimate criminal investigation."*

For no minority can for long rule an actively hostile majority. It is said under modern conditions of destructive weaponry, a minority can tyrannize permanently over a majority. This statement ignores the fact that these weapons can be held by the majority, or the agents of the minority can mutiny.

Mutiny: refuse to obey the order of a person in authority. An open rebellion against the proper authorities.

I am speaking to the all agents in question.

As David Humes' essay "Moral and Political" states: *"Nothing appears more surprising… than the easiness with which the many are governed by the few and the implicit submission with which men resign their own sentiments and passions to those of their rulers. When we inquire by what means this wonder is effected, we shall find that because force is always on the side of the governed, the governors have nothing to support them but opinion. It is, therefore, on opinion that government is founded; and this maxim extends to the most despotic and most military government."*

You FBI agents must extract your head from your ass before your destruction occurs. I know there are good people working at the FBI and are disgusted by the actions of their income provider. You'd better get your mind in line with freedom and unalienable rights, last chance.

The insurrection was no different than Black Lives Matter and ANTIFA throwing firebombs at federal buildings. Most all Black Lives Matter and AN-

TIFA were fined and released. It is said the people who were let into the capital are being held in solitary confinement. Their only fault is they weren't paid by leftist billionaires who conspired with elected officials, instead they were psychologically directed by Qanon.

We need not to investigate the National Security state, their actions have provided all the evidence we need. The FBI, CIA, Department of Defense, Department of Justice, and many more agencies have over stepped their boundaries and must have their fears exploited.

The Military is mostly full of Americans who signed up to lay their life down for freedom. Soldiers are patriots, and believe in the oath they took to defend the Constitution. We must destroy the dark forces that wish to imprison and destroy our freedom. It is likely we can trust most all from the First Sergeant down, but must be aware of our surroundings at all times. We will be sparing soldiers' lives by destroying this dark force called Communism, we will be saving our best and strongest for the real fight of good and evil.

Everyone is falling for the political circus of pulling out of Afghanistan. Not only did the Afghans lay down their weapons when the Taliban arrived, they also misappropriated billions of American tax dollars, they were draining us. The Afghans continued on the humanitarian tit and didn't do anything to set themselves up for long term survival.

All the people I knew that worked in Afghanistan knew far ahead of time when the last plane was leaving. If they didn't get on a plane to America, they probably shouldn't be on a plane to America. If you want to go after Biden, do it with truth. Anything less will prevent our success. Afghanistan had nothing to do with 9/11, the invasion of Afghanistan is some George Bush bullshit. We need not to guard the opium fields anymore because Trump busted the opioid rings.

The military is a weapon designed to kill our enemies brutally. We must rethink military of purposes of college. We need killers in the military, we must eliminate from the ranks anyone who can kill not. We will prohibit foreigners from entering our military or in operations within all US and state government.

Saddam Hussein was trained at Lackland Air Force Base in San Antonio, Texas in the 1964-65. Saddam belonged to us and when we didn't need him anymore, we killed him. I care not for his life, I'm concerned our government is indeed selling us the problem and cure as they have on many occasions. These foreigners take the training, information, and connections then use it to exploit their home countries, like the cartel in Mexico.

We must have a heavy organized civilian militia presence in Texas. Operational and ready to stand against federal entities like the FBI, CIA, ATF, Homeland Security, TSA, National Guard, even the Texas State Guard (who is under the direction of the governor), even school boards if necessary. We must have an organized militia to ensure our unalienable rights.

If local governments (mayors, city councils, police, judges) violate the constitution they must have their liberties arrested and suffer consequences. All said entities' purposes are to ensure Americans' unalienable rights, all said entities are designed not to protect us.

Law is reason without emotion—without emotion you are no longer human. It is good to control your emotions, but without emotions you lack feeling. Feeling is to be human, when people lose their humanity, I lose humanity for them.

Rebellion indeed induces the search and acceptance of darkness. We want to rebel from a hidden darkness and get caught in darkness. The spirit of rage is radiating all over the world. This rage comes from a feeling of injustice, we must realign the minds of our institutions. There is only two things that a psychopath fears—the loss of utilities and death. We must provide both to truly ensure success in freedom and the future of our nation.

The cause of America is in great measure the cause of all mankind. - Thomas Paine

The Corona/Covid - 19 virus has made everyone forget Jeffery Epstein and all accomplices and clients. Ghislaine Maxwell and many more have the answers to our questions, and who knows what they are doing to her to ensure her silence. We must know the flight list.

If there isn't a lot of high profile convictions from this trial, there will be consequence. It can be for the perpetrators or the judges and the Justice Department can carry the lions share. For clarification—a debt is owed for crimes in this Universe, aiding and abetting will cause you to take their place. They will convict this stupid bitch—but all the predator men will go free.

Ghislaine Maxwell isn't the a scapegoat—she's the golden goose—she better shit some gold before she mysteriously hangs herself like Epstein.

Everyone has taken so many antibiotics to a point where they don't work anymore. We need not to edit our genes, our genes are fine without destructive pollutants mutating them. We need not to meld or load our consciousness into a computer. A computer has not the capacity that human have. There are many ways to redirect your consciousness. Knowing Universal truths, like the Constitution, can help clear the clutter.

We all have rights documented in the Constitution of the United States to do as we choose with our bodies, that includes abortions and vaccines. Most abortions consist of taking a pill to produce hormones that excrete the pregnancy. But I ask you, who are Americans protecting by stopping abortions?

Former KGB Propagandist Defector Yuri Bezmenov says it takes 10-15 years to demoralize a nation, which is complete and to the fullest extent in America. America has been exposed to the enemies "Ideological Subversion" (Marxist-Lennonist) with pressure without challenge for more than three generations. The indoctrination is complete and irreversible, new facts are unacceptable and most will refuse to believe their own eyes. The only way to counter this brainwashing is to produce a generation of patriots acting in favor for America. The influence of Marxist-Lennonist ideology has already influenced our defense and economy. The last two measures are crisis and normalization.

Former KGB Propagandist Defector Yuri Bezmenov explained that the psychological conditioning is complete and irreversible, and we must produce a generation of patriots. So I ask you again, who are you protecting by stopping abortions? Just stop for a minute and think of who is having abortions.

Yuri Bezmenov said America is the last free nation, if the boat goes down, so will world freedom. Bezmenov also said to stop paying communist governments, millions of Soviets will thank you.

If a women wants an abortion then she should be able to have one. By no means will we allow the material excreted by the abortion be used for science or to sell for profit on a legitimate or black market. The material will be incinerated. Anyone who complies not will have steep penalties applied to their lives. Then we will see if abortion rights are still a hot topic.

The truth is, abortion or focusing on Afghanistan is controlled opposition designed to keep your attention elsewhere. We are being killed by world wide pollution and dangerous frequencies, and will continue to be killed to prevent an irreversible world wide economical collapse. We can survive not the pollution and frequencies, what difference would it make if we went broke? What good is money to a deadman?

Agenda 21 blueprint is to inventory and control all land, water, plants, animals, movements, production, construction, energy, information, and all humans in the world.

It's the biggest public relations scam in the world, and is implemented locally. It's designed not to salvage the land, water, plants, animals, movements, production, construction, energy, information, and all humans in the world—its design is to own all said entities by the government. It works not in Communist China, what makes you think it'll work anywhere else? The difference between Communist China and Americans is weaponry and the desire to use it.

We will limit construction and production in areas of condensed consumers and wasters, where the land, water, and air is overpowered by pollution and over consumption of natural resources. We must live within the real limitations the Earth provides.

We must fine tune our energy source and discard outdated technology where the dangerous by-product far outweighs the benefit like coal, oil, and nuclear reactors. Creating energy must be a human effort, using Capitalism and slavery to produce energy must be abandoned. We must use the existing technologies that are already present, unknown, and effortless in creating energy.

We are using dangerous Smart meters to measure water and electricity usage, all the while these service companies that provide us with conveniences are harming us with harmful microwaves to lower cost. Californians say Smart meters can burn down any house the operators choose, and most wildfires start from Smart meters. Both the water and electric companies can only profit 10% above their operating cost. They waste the money they saved at the end of the year on trucks, equipment or what ever they have to do to get rid of the money they have in surplus.

Both said entities can easily pay meter readers so we can be free of these dangerous devices. This tells me that the meters have another purpose, a hidden design to spy on consumers in this surveillance state we live in. The cost to the consumer is harmful radio waves and this far outweighs the benefit and the invasion of privacy.

They are simply cutting out paid humans for robotic slaves, the enemies of freedoms' main focus is to eliminate American jobs. The word robot is Slavic and means slave. Can you be in competition with free? This is all handed down to the corporate psychopath from their communist owners. Anyone that partners with the communists are the enemy of all living beings and must be punished accordingly.

The idea is to make public lands off limits. Having everyone live in high density housing to make it easier to control and surveil humans in a global totalitarian state, it will destroy our land and resources. They would like to move people from the rural lands and would refer to them as the wild lands, but they can like in one hand and shit in the other and see which one fills up the fastest.

It's a stealth plan and social engineering disguised as humanitarian efforts. Individual rights outlined in the Constitution of the United States outweigh the cause for the greater good and the rights of the community, we learned this from the "Bystander at the switch."

The 1948 United Nations adoption of the Universal Declaration of Human Rights after the Second World War due to the human rights violations committed during this period.

It explicitly forbids governments from treating some humans more valuable than others or from sacrificing some for the benefit of others.

It also forbids governments from knowingly imposing harm on some individuals in order to serve an alleged "greater good," and forbids governments from imposing a hierarchy of rights on their citizens. Had it not been for this, NAZIs (National Socialist Zionist) would still be in control.

The communist faction in our government, along with the media, produce a continual heavy crisis to control people who will eagerly give up their individual rights for the greater good. We must redirect these people who continually agress on others because of masks, for or against. Shut your fucking mouth before the pendulum swing, I say this to both sides. Mind your own business before you get physically aggressed upon. The action you are committing is harassment and will be tolerated not. You idiots are easily divided.

There are humans claiming they need to refuge in our country, but tell me after they arrive they are here because the money is good. If asylum seekers want to refuge then they should be able to. Emitting them into our workforce is taking from one to give to another—that is insane, against Universal Law, and human rights law listed above.

They need to stay in refugee camps until refugee camps are over filled and can't be sustained by tax payers—until we all have had a gut full of pain. Then we can gather enough refugees to lead a physical attack on their oppressors.

If there are no oppressors then they need not asylum. Continuing an endless cycle of workforce invaders is over and done.

The only American citizens that want immigrants to enter into America are people who work in the media, realtors, people on welfare, the academic sector, and obviously companies who stand to gain from cheap exploited labor like grocery stores and construction companies. If they approve of the idea then we should let only those immigrants that preform said jobs entrance—minus grocery stores and construction companies. I think foreign professors, news anchors, directors, cameramen, and realtors will make a great addition to our roundedness of the world.

We can vote on the matter and you can state your profession. We can let one immigrant in for every profession of a natural born citizen that votes for entrance. If you are a bureaucrat, then we will let you compete with one immigrant who is a bureaucrat. If you have not a job, then you vote not. If you don't at least come from a place that has freedom of speech then you're welcome not.

Matter of fact, we need to invade the land South of Texas. If they leave an open spot in their community we will take it up. They love the American way, lets give it to them. When you do move into their position, do as the Romans do. Price them not out of their homes, raise not their taxes, build not modern monstrosities, force them not to think as you, over populate them not. Just live as they do or stay home.

I know and have been in relationships with immigrants, there are good people everywhere you go. Many people in my family are Mexican, from both sides of the border. There are immigrants in America I am thankful to know. Capitalism and the real limitations of the Earth in condensed areas are a fact that is undeniable.

I enjoy people from other cultures, especially people south of Texas. I grew up in Tejano Days at the Houston Livestock Show and Rodeo; I grew up in an area full of illegal immigrants. There is nothing more Texan than Mexicans. But we have our own Mexicans that didn't cross the border, the border crossed them. They have lived in Texas before it was Texas. Restricting entrance into America has nothing to do with xenophobia, it has to do with preserving life giving resources and restricting destruction from over population. Use your eyes and your brain, think for yourself.

Anyone who is aiding and abetting illegal immigrants by giving them airline tickets and sending them deep into America or introduces Afghans into our workforce will face harsh penalties. If they are refugees because you destroyed their country, they will stay in refugee camps until further notice. Failing to do so will result in harsh penalties.

If we in America are the beacon of light—let's set the pace simpler. We need not technologies that occupy the mind and soul. We have existed for millennia with out it—we must discard it before we kill us and all the birds, bees, flowers, and the trees with destructive frequencies and poisonous air. We are lost without the living beings that inhabit the planet with us. We need them as much as they need us.

Fret not—we will be moving into something wonderful. We will cease in spending our every waking moments trying to create energy or being a tick on the dogs ass that creates energy with harmful consequences.

It's that new thing that all will welcome and will be voluntary. We are the true co-creator with the source that provides for all and we will be able to live within the limitations the Earth provides. There are wonderful things that benefit all living beings, things we could have never imagined. They will come. Things that destroy not the earth, food, water, air, mind, body, soul. We must have discernment from imposters.

Nikola Tesla proved he could tap into the energy of the Ether and proved it is the cause of every magnetic field, electricity, and all energy—electricity couldn't exist without Ether. But like Antaeus, we derive our power from Earth and the oil insulation. Tesla knew nothing of biology or how to draw, so "attaching our machinery to the very wheel work of nature" may not be the best way. We must have the wisdom to stop when injure occurs.

In the story of the Ark of the Covenant, Uzzah touched the Ark and was struck down by God. Scientist John Hutchinson and a team of experts built a replica of the Ark of the Covenant. With limited instructions from the Bible they created a technological device capable of an output of 50,000 volts of electricity.

There is nothing new about humans and we have had amazing technologies for thousands of years. Prehistoric man has documented time and space travel with levitational devices that defy the laws of physics. There are hieroglyphics in the pyramid and on cave walls of lightbulbs, tanks, helicopters, cellphones, and a great deal more devices we use today and technologies that are to come.

Just as a perfect pyramid provides energy, from dimension alone, other dimensions in our existence can draw energy from the atmosphere. Hutchinson created levitational devices as well, just think of the

UFO's and how they operate. Hutchinson isn't the first to understand this technology, there are beings that have solved our physics problems and move effortlessly through our atmosphere. These hovering and traveling devices appear to not emit exhaust.

If these vehicles that levitate and defy the laws of physics as we know it are indeed harmless to all living beings, then we will soon utilize said vehicles. We must have discernment of the input of material for energy and the output of the modulation, it must be minimal if not completely non-existent. One reason we haven't shot one down is because they are more advanced than us. We must be brave, they are already in violation of Universal Law and are hiding because they have something to hide. Let's get a few, if the technology already exists we might as well use it or rule it out.

We will cease crawling across the ground like a lizard and laugh at our crumbling roads because we will need them not. Many people today can live minimally, it's possible that these levitating vehicles are the new houseboats on the bayou. You can move where you desire without deforestation or beehive living. It will allow humans to spread out and live in nature without the destruction of one to give to another.

The American Bison (Cibolo) is a prehistoric animal like the Gator. American Bison eat like deer (grasses, weeds, and leafy plants) and produce more meat per consumption than cattle. Meaning there is less need for deforestation for cattle fields. American Bison selectively eat grass contributing to the bio-diversity of the earth and love to travel behind wildfires and eat new vegetation. This habit contributes to the fertilization and replanting of North America. American Bison will travel from Mexico to Canada if permitted.

Capitalism and the genocide of the Native American was the driving force in the extermination of an easy food source that can eat like a deer. Replacing the American Bison with cattle caused there to be fences and constant work in maintaining cattle. In the devilish means of capitalism, we were forced and conditioned to substitute a free product and replace it with more work with less return, it's hard to sell a free product. We need not destroy cattle operations, we just need to replenish Buffalo.

Buffalo clover (Blue Bonnets, the Texas State flower) are poisonous to most all other animals excluding American Bison and deer.

With the replenishing of the American Bison we can have a food source that has been enjoyed for millennia. Texas has the largest whitetail herd for a reason, to ensure survival. What good are they if they die from polluted water, air, and over development of America? It takes four or five deer to amount to one hog, and who knows how many deer equal one American Bison. I'd take a hog over deer any day. Wild hogs are not native to America and considered a nuisance because they destroy property, this is city folk mentality. Wild hogs and the American Bison can provide for all, and no one should go hungry.

A women in Kenya built a factory that turns plastic waste into bricks that are stronger than concrete. It's a good way keep the tiny particles of plastic out of the ocean, but just the fact that petroleum plastics are made is too harmful and dangerous to be in the hands of humans.

Hemp plastic is non-toxic and biodegradable. There are many more types of plastic we will use that are harmless and useful. We have technologies that remove plastic completely from our lives. We have products that are beneficial to the environment that preserve and contain. We have ways and will use them.

The homeless are used as psychological weapons to drive prices down on real-estate and cause dissonance. In the 2009 movie, The Education, an older Jewish man courts a sixteen year old girl with flattery and the appearance of wealth in the 1960s. Even at sixteen years, the girl sees through the tactics of the older Jewish man.

The older Jewish man buys properties and houses in neighborhoods he sees promise, properties that old people own in neighborhoods that are forgotten. He then moves black people into these neighborhoods, giving the older 1960s residents who at the time were most likely racist a feeling of fear they will be robbed or their property value will drop. The 1960s residents quickly move and he buys the properties cheap.

This is what is happening in Austin, Texas. Mayor Adler, along with his wife, appear to own around $300 million in real-estate. Mayor Adler is most likely the front man for a larger conglomerate of outsiders who care not for the homeless, humans, or Austin. Greed and money is the agenda. Most humanitarian efforts today are a disguise for greed and power.

They claim they want to help the homeless and give them benefits, chip them like cattle, and scare the hard working residents from their homes with mentally-ill pawns. Giving them money and a place to stay draws more humans that have abandoned civilization and the rules of civil living apply not. Meaning they shit on the sidewalks in plain view and no one said anything about it. The rules need not apply because there is no punishment available. You can't take away their liberties or property, because they value it not.

People who can't think for themselves are always directed and used like cattle. All humans will work by the sweat of their brow, we are all drawers of water and cutters of wood. There is no way around work to survive in this reality.

A better way to help these humans is to give them purpose. If no one has loved them and they don't love themselves, then they will just exist, act out for attention, and let their meat vehicle and animal instincts overcome their souls that should be making decisions for their bodies. Most all the homeless I know have all been abused, and injected harmful and powerful drugs that have altered their cognitive ability maybe permanently.

If their souls have abandoned their body then why should the body continue? Give them what they desire, harmful and powerful drugs until their bodies exist not. Contain them while they self-destruct. No point in drawing the inevitable out, every human has the right to do as they choose with their belongings, including their bodies. It's a harsh and radical idea, but we can continue not with this millstone tied around our neck.

You can help not someone who wants not help, their rock bottom isn't your rock bottom, and no-one can stop this way of existing until they have truly hit rock bottom. They want you to watch them die, they want someone to have an emotion that resembles tragic love towards them. They will continue to live in your view to cause social discourse, vampiricly thieving your consciousness and resources. Mothers, recognize this in your children and correct it with real love.

For those Texans that are just in hard times and had a series of events that sent them into a downward spiral, we must lift them up. We must lift up anyone who really wants help.

The 1939 Works Progress Administration or Work Projects Administration was an American New Deal agency. They employed millions of humans to carry out public works projects. These humans were employed to build public buildings and roads.

This project created jobs that served the public good and conserve the skills and the self-esteem of workers throughout the U.S. The WPA employed out of work laborers, along with artists, writers, and musicians.

Congress approved the Emergency Relief Appropriation Act of 1935, this was the relief bill that funded the Works Progress Administration. This program was created by President Franklin Roosevelt to relieve the economic hardship of the Great Depression that Woodrow Wilson created by implementing the Federal Reserve.

The public project was disbanded in 1943, World War II was a great way to stimulate the economy and war is always a racket. These people performed public works projects such as creating parks, and building roads, bridges, schools, and other public structures.

The homeless that want to climb from their pit should be able to do so with the help of a program like this. It will take from private enterprise, but if it's public works we should do what we can to help the public in taxes and employment. The roads in Beaumont and Orange, Texas have been in construction for 20 plus years. Many deaths have occurred because they can't seem to finish the project with private enterprise. Let's cut our cost and employ those willing to work with the guidance of skilled craftsmen. An apprenticeship could give new skills to our naturally born citizens.

The saying during this public project time was "Two a diggin'—two a hoein', two a crapin'—two a mowin'." At Tyrrell Park in Beaumont, Texas a structure is being refurbished, this structure is a Roosevelt building—built by people from the WPA.

We have an immediate public work needed. The State of Texas is overrun with invasive plant species, one of which is the Cedar tree. This tree consumes a massive amount of water, and is a wonderful fence post. We can employ the homeless to eradicate the Cedar from Texas and sell fence post. This will create jobs that are beneficial while producing income.

Any public land or voluntary land owner that wants Cedars removed can have this done for the preservation of water and allowing the state to reclaim and sell the fence post. This is just one of the many things that can be done. This can potentially give the homeless the opportunity to relocate to a place that they can afford.

I lived and worked in Austin, Texas and damn near went broke because the cost of living was higher than a Giraffe's ass. We must revert back to what Doug Sahm and Waylon Jennings said, "You just can't live in Texas, unless you got a lot of soul." It's all white bread with high income from tech handed down from their communist and Saudi owners who become rich and have unlimited amounts of money through state sponsored murder, theft, and slavery to dispute and disrupt capitalism.

All Hospitals and Insurance companies prior to former president Richard Nixon were non-profit like the electrical and water companies. Richard Nixon along with Henry Kissinger aided the Communist Party beginning in 1972 and aided the already existing development of communism in America through corporations. These two ignorant wretches thought they were helping capitalism, all the while the communists were controlling them to destroy capitalism. Or they were easily paid off to imprison their brother.

We need not socialism of any kind, we have had good systems that have been tainted by demons like the Communist Party, Richard Nixon, and Henry Kissinger. Insurance is a pool for security, hospitals are for the purposes of a humane existence. Neither should depend on the other and both will be non-profit.

We have undergone trials and have experienced illuminating revelations. The year 2020 indeed gave us clear vision of the plan that was in effect for probably more than one hundred years. The achievements and tribulations that we received are the manifestation of our character, and are the ones we are ready for.

We are all operating in relation within a system of economical servitude, conditioned media, social expectations, and prescribed indoctrination. This system will release us not of our true humanity. We will live within systems as human beings triumphant and then destroy the needless. We will listen to the spiritual demands of our lives and resist programming to prevent delusional schizophrenic breaks.

Aligning ourselves with the programmatic life requirement is making an enemy of our mind, body, and soul. We must listen to ourselves and evoke our higher nature rather than our lower.

Fear in you is protecting your ego and holding you captive—what we want, what we believe, what we can do, what we think we love, what we regard as the aim of our lives. These fears are too small and hold us down to the ground by the throat. If it's simply doing what the mass society tells us to do then it is holding us down. Peer pressure and uncertainty is holding you captive.

Following our bliss and being brave enough to follow our bliss is the solution to that which hold us down. Shifting the same world around and changing rules will save not the world. Living our lives and bringing our lives to life will. Being centered is being compelled not by desire, fear, or by social requirements, it is your center and yours alone. We are the consciousness of the Earth, we are the Earth.

Why should the Auschwitz concentration camp remain standing? As a reminder of murder and torture? The reminders of our past have been removed to initiate the renewal, but will fail.

The spirit world is the guide to our deep inner struggles, inner mysteries, inner thresholds, passage. If you have not the spirit world you have to work it out for yourself, time passes quickly. The spirit world and experience from your elders can give you clues early in life to what is to come and help remedy before you're too old to do so, it can remedy regret. Regret not, the spirit world can guide you and give clues to your existence.

Have the experience of being alive, feel the Rapture of being alive. You must participate in the game of life, the here and now is the time to have the experience of life.

There was a time when the animals taught us how to survive in nature, back when man was a newcomer. We owe the animals a debt for teaching us the correct way of living. Rituals for thanks were given to these beings, whose meat gives us life, whose nature showed us how—messengers. Rituals of recognition of our dependency, respect, and in doing so we perform the work of nature. They are messengers from the other world, the ritual is an enactment of a myth and nature.

The Bible does say to be fruitful and multiply, but it also says to replenish the earth. The Buffalo dance is to replenish the buffalo. The destruction of the buffalo was a sacramental violation.

Over the years, groups have taken the reigns of acceptable truth for control. The Hebrew started the subjugation of women and goddesses. The Hebrew come from a people of fighting nomadic hunters that turned into herding people of the father. Most all other and older religions revere women.

The purposeful misinterpretation of the Garden of Eden is the beginning of duality with male domination. We do have roles, but one can exist not without the other. You must pass through a woman's legs to enter into this reality.

Just as we let the child die away from us to become adults, we left the Garden of Eden to return with Jesus, the going and coming we must all experience. The Tree of Immortal Life is the return to the garden, to return to our spiritual body, the fruit of the spirit. Knowledge of good and evil expelled them causing an experience of duality from everything. Duality sent them out to experience separation. We must acknowledge our spirit and return as one.

We and the father are one, death of the flesh is the birth to the spirit. Your soul possesses a body, when you die it becomes the spirit. As the circle you start, so shall you end—in and out of pastures. No tongue has befouled or can reach the name of God. The mind can understand not, it must be felt within, you must ask.

The Constitution of the United States, the Magna Carta, the Code of Hammurabi, the 10 Commandments, all express and document rights given to us by a higher power. A machine has none, without these rights we are reduced to the level of a machine, and have elevated a machine above us. Robot is a Slavic word for slave.

We are living in a Ray Bradbury Theater movie called "The Murderer." Maybe there will be a rise in Diathermy retaliation. We must rise up as the troubadours of love and freedom, and watch the men of action fall back.

     Not all police, judges, clergy, pipeline/refinery workers, Saudis, or anyone else I mentioned are bad. I have had many men who claim to be Masons help me and even tried to retrieve the information I write and didn't understand or remember. They fiercely wanted to know who these demons were, but I understood not. But as long as we want to be free, we are all innocent until proven guilty.

     The corporate communists have shown their hand and haven't anymore cards. They are all in and we're just getting started, like Ravel's "Boléro." I say death before dishonor. I would rather die a warrior's death than be poisoned and burnt out of existence like a rat. You will be able to conquer not the world because it belong to the unmoved mover moving in all that moves.

     We must eradicate the people and pollution from Africa that pollute the entire earth or we will not survive long enough to correct anything else—the summer of 2022 will be deadly. Once we correct Africa (the largest source of pollution on the planet) and the UN spraying chemicals on all humans, then we will work our way to Russia.

We must have a standing civilian, well organized militia. We need not training—we've run down wild boar and killed them with knives and spears all of our lives. We must agree on what we can stand and what we can not—and then agree what the consequences are and apply.

There has been many attempts to test, disrupt, and discredit my memory—which tells me that many are in fear of it, and should be. There has been many attempts on my life including sabotage of vehicles and attempts to enrage me or others to make a bad decision.

If I have risked my life for you and saved you from the deep and the pit of hell then you better check the Karma box. I'm not asking you—I'm telling you to see, understand, and act.

Nathaniel Welch

August 29, 2021

Additional Sources

Thomas Paine "Common Sense"

Timothy Freke & Peter Gandy "The Hermetica"

Joseph Campbell

F.E. Peters

Dr. Thomas Cowan MD

Murray N. Rothbard "Power and Market"

Billy Carson "Compendium Of the Emerald Tablets"

Stephan Molyneux

Democide - R.J. Rummel

Dr. Alan Greer

FRANKENSKIES

Gene Roddenberry

Leonard Cohen

Peter Rowan

Nome Chomsky "Manufacturing Consent"

Cover art and First Nations art by: Michael Snowden

Instagram - @mchlsnwdn

Behind the Fence Photo by: Randy Edwards -

www.iamrandyedwards.com

Train Wreck Photo by: Robert Arabie